Scale-Free Networks

Scale-Free Networks

Complex Webs in Nature and Technology

GUIDO CALDARELLI

Statistical Mechanics and Complexity Centre INFM-CNR,
Department of Physics, University of Rome 'Sapienza'

OXFORD
UNIVERSITY PRESS

OXFORD

UNIVERSITY PRESS

Great Clarendon Street, Oxford OX2 6DP

Oxford University Press is a department of the University of Oxford.
It furthers the University's objective of excellence in research, scholarship,
and education by publishing worldwide in

Oxford New York

Auckland Cape Town Dar es Salaam Hong Kong Karachi
Kuala Lumpur Madrid Melbourne Mexico City Nairobi
New Delhi Shanghai Taipei Toronto

With offices in

Argentina Austria Brazil Chile Czech Republic France Greece
Guatemala Hungary Italy Japan Poland Portugal Singapore
South Korea Switzerland Thailand Turkey Ukraine Vietnam

Oxford is a registered trade mark of Oxford University Press
in the UK and in certain other countries

Published in the United States
by Oxford University Press Inc., New York

© Oxford University Press 2007

British Library Cataloguing in Publication Data

Data available

Library of Congress Cataloging in Publication Data

Data available

Printed in Great Britain
on acid-free paper by
Antony Rowe, Chippenham

ISBN 978–0–19–921151–7

10 9 8 7 6 5 4 3 2 1

Σὰ βγεῖς στὸν πηγαιμὸ γιὰ τὴν Ἰθάκη,
νὰ εὔχεσαι νᾶναι μακρὺς ὁ δρόμος,
γεμάτος περιπέτειες, γεμάτος γνώσεις.
As you set out for Ithaka,
hope the way is a long one,
full of adventure, full of knowledge.
C. Kavafis, *Ithaka*

Love is a temple, love the *power* law.
(adapted from)
U2 *One, Achtung Baby*

Contents

Preface

Midway upon the journey of our life, I entered the field of scale-free networks. This topic was completely new, at least for me. Neither guides nor textbooks were available for an overview of the basic notions. The purpose of this book is to become a useful reference for students of various disciplines. To this wide audience I try to report the occurrence of and the reasons for the ubiquitous presence of networks in biology, physics, economics, computer, and natural sciences.

In this volume I have collected some of the most interesting case studies in these various fields. The aim of the book is to give a unified theoretical description for these various cases. Whenever possible, theoretical results and models are described by reducing the formalism to the minimum. This book is specifically shaped for readers who 'already know something' and are curious about the rest of the story. An example is that of students who wish to begin a PhD project on networks (regardless of their background). I assume that they have a basic knowledge in either computer science, or statistical physics or biology, and they need some introduction to ideas and results in the other fields. Another example could be an expert in techno-logical networks who wants to discover how similar ideas apply to food webs and so on. The ideal audience is therefore made of researchers who want to start the study of the scale-free networks and need a reference textbook to establish the basic ideas about the field.

In my opinion, this presentation of the various applications is particularly important, since the application of graph theory is really an interdiscipli-nary science, where most of the results derive from dramatically different situations. Maybe, despite my efforts, a little bias is still present since my background is in statistical physics. Possibly more emphasis is given to the study of statistical distributions than to other quantities. I hope to convince my readers that this is indeed a key point rather than a bias. The idea is that a statistical approach is unavoidable since networks are very large sys-tems, whose collective behaviour cannot be understood from the elementary features.

Let us consider an example very familiar to readers with a background in physics. In thermodynamics a system is usually composed of a huge number of particles. A standard approach would be to study the behaviour of every single particle and see what happens. The problem is that the number of particles is of the order of the 'Avogadro number' a fantastic $602,300,000,000,000,000,000,000$ which no present computer would be able to handle easily. Statistics does not only solve the problem of processing this information, it also gives an insight to what is happening in the system. If we take a box full of gas, the only thing we can do is to study the average motion of the particles inside. This average is not only a way to process data. Rather the average velocity of the particles that collide with the boundaries contributes to the macroscopic quantity that we call pressure. Exactly in this spirit a whole series of physical quantities like energy, temperature and pressure has its origin and definition in the statistical description of the system. Remarkably this is exactly the situation for scale-free networks, as we shall see in this book. In the World Wide Web millions of pages are added to the system every day. This microscopic 'motion' produces global properties like community structure, PageRank, global resistance to failure, etc.

The challenge of this book is to keep this statistical approach while still being accessible to people with little or no background in mathematics, statistics, or physics. To stay in between a rigorous derivation and a simplified version is a serious challenge since it could cause a double disaster. From one side, sometimes it is impossible to simplify the mathematics too much. This means that, regardless of the efforts, the exposition remains obscure to some of the readers. From the other side, the 'experts' in the field might consider all this simplified exposition rather boring and dispersive.

To avoid such dangers, I have given this book a particular structure. Formal derivations (unnecessary but useful), proper definitions, and other notions are all reported in the appendices. For a network textbook some formulas are necessary and unavoidable, others can be consulted or not. In general, I give a suggestion in the latter case; extra paragraphs are indented and indicated by increasing signs of difficulty:[1] ☕(during a break), 🍷(some concern), ☢(geek only).

☕ The average of a quantity is defined as

$$\langle k \rangle = \int kP(k)dk.$$

Readers are encouraged to skip whatever they believe to be an unnecessary

---

[1] I am indebted to T. Fink for such an effective nomenclature (Fink, 2006).

explanation and focus on the topic of interest. Whenever possible, chapters have been written without explicit reference to previous parts. The idea is that readers with different backgrounds can find their optimal path through the notions of the book. The various chapters are as far as possible independent entities for which we try to give the largest possible bibliography. This should allow readers to find the pertinent manuscripts or textbooks. To help the reader an electronic copy of the bibliography is also available at http://www.citeulike.org/user/gcalda/tag/book. From this site it is possible to export the bibliography or consult (subscriptions might be required) the various papers. For absolute beginners in a particular field we put at the beginning of every chapter a synthetic view of the basic notions.

> Similarly to what has been done for the formulas, we have condensated for pedagogic reasons the abstract of every chapter in a shadowed box. These boxes should not be skipped!

A particular remark is necessary concerning the appendices dealing with basic mathematics and statistics. I assume that basic mathematical concepts (like matrix analysis, probability, frequency distribution, or average quantities) have been properly defined to the readers in some undergraduate course. Appendices report only a brief list of the quantities used in the book just to refresh the memory. In any case, these appendices do not represent a substitute for a good mathematics handbook for those who need it (every now and then, we all need it!). As further help, I also include a summary of the key concepts and a glossary of the terms used through this book. Finally, whilst all efforts have been made to avoid errors, it is possible that some may become apparent at a later date. Please check the website http://www.scale-freenetworks.com for a list (if any!) of errata discovered after publication.

At the end of this preface I want to cite some persons who have had a role in the writing of this book. It is a pleasure to thank S. Ahnert, A.-L. Barabási, A. Barrat, G. Bianconi, K. Börner, A. Capocci, F. Coccetti, P. De Los Rios, G. De Masi, T. Di Matteo, T. Fink, A. Gabrielli, D. Garlaschelli, S. Leonardi, A. Maritan, Z. Nikoloski, P. Paci, L. Pietronero, F. Rao, G.P. Rossini, V.D.P. Servedio, R. Setola, A. Vespignani, and G. Weisbuch for their comments and suggestions. Many colleagues also shared with me their results, providing me data and figures. Thanks to S. Battiston, C. Caretta Cartozo, M. Catanzaro, B. Cheswick, D. Donato, H. Jeong, J. Kertész, F. Liljeros, N. Martinez, M.E.J. Newman, F. Rao, E. Ravasz, D. Stouffer, M. Styvers, J. Tenenbaum, M. Thrasher, P. Tosi, and A. Vázquez, for sharing their work with me and for allowing me to present in the best way the contents of this book. For the same reasons I also thank the authors of the free software Pajek which I used to visualize most of the graphs shown here.

One of the most difficult things when you write is to face the white paper in front of you. A good start is certainly a big help in this situation. Many friends gave me some hints to start the various chapters. Amongst them I would like to thank Mikhail Afanasievich and John Ronald Reuel.

It has been quite hard for me to write this book. Many times I felt *all thin, sort of stretched [...] like butter that has been scraped over too much bread. (I wanted) [...] to find somewhere where I can rest. In peace and quiet, [...] somewhere where I can finish my book.* Thanks to the city of Paris and the École Normale Supérieure for providing such a place. Thanks to Sonke Adlung for being an editor with plenty of patience. Thanks to Chloe Plummer and Julie Harris for helping me so effectively with all the problems I had with the manuscript. Thanks to the many friends who reminded me that 'et facere et pati fortia Romanum est' and finally to Raffi, Andrea, and Laura for their inexhaustible support.

Rome,
April 25th 2007

Guido Caldarelli

Introduction

Alice was beginning to get very tired of sitting by her sister on the bank, and of having nothing to do: once or twice she had peeped into the book her sister was reading, but it had no pictures or conversations in it, 'and what is the use of a book,' thought Alice 'without pictures or conversation?' (Carroll, 1865). This book is about self-similar networks and it wants to be of some use, therefore I decided to put on top of some conversation all the pictures, diagrams, and plots needed to explain our topic properly.

In this book we present the theory and experimental data available for scale-free networks. Examples range from the structure of the Internet and that of the WWW (we shall see in the following that they are different systems) to the interconnections between financial agents or species predation in ecological food webs. Thanks to the simplicity of graph theory it is very easy to provide a network description for different systems. Network components can describe many different real-world units such as Internet providers, electricity providers, economical agents, ecological species, etc. The links between the various components can describe a global behaviour such as the Internet traffic, electricity supply service, market trend, environmental resources depletion, etc. It is clear that the shape of a network and its functionality must be closely related. This means that if we know the topological properties of a food web, this could help to determine the laws governing predations in ecosystems. In principle, we could discover and protect the key species (if any) that form the skeleton of the ecosystem. In a different field, the same theory can help in designing a faster and more efficient Internet network, avoiding possible attacks on its functionality. If it is easy to define a network in almost any field of research, there is no reason why different networks should have a similar behaviour. Yet, from experimental studies, we find that almost all the networks considered here (and many others) share many similar properties (but also maintain many differences). One possible explanation for this universality is that a common formation mechanism acts in different cases. If the nature of this mechanism is understood, this piece of information could be used to predict the future evolution of such objects. Other explanations are possible, but I dare say that we lack a general understanding of this phenomenon.

As already mentioned in the preface, all these systems are formed by many interacting components whose collective behaviour cannot be predicted in terms of the components. Consider that in some cases, not only the global structure (e.g. the cable connection of the Internet) but also the dynamical evolution of the system (e.g. the Internet traffic) are the self-organized result of the interactions between network elements. For the above reasons such systems are a paramount example of the so-called science of 'complexity' whose presence is becoming more and more evident in physics, biology, mathematics, and computer science. Both in scale-free networks and in other complex systems the presence of large-scale correlations is witnessed by the appearance of power law distributions. It seems then natural to spend some part of this book in the description of these particular statistical distributions. Sometimes, as in the case of fractals, these power laws appear in nature and describe geometrical properties (e.g. a wildfire that leaves unburned regions). In other cases, we have natural phenomena whose geometrical properties are regular but whose evolution over time proceeds through power law distributed avalanches (e.g. species extinctions in an ecological system). In scale-free networks (and this is the core of the book) the power laws appear when considering 'topological' quantities such as the degree (defined as the number of edges per vertex). This does not imply that scale-free networks are simply another type of fractals. Rather, the scale-free nature of some real networks might have the same origin as the scale-invariance present in other phenomena. One example could be that of the multiplicative processes that can produce both power laws and log-normal distributions (these can appear as power laws). A similar mechanism can then produce fractals in one case and scale-free networks in another. This common origin explains why in most of the cases scale-free networks present a series of properties usually associated with self-organization and complexity. They effectively start from small collections of vertices and edges and in their growth, they develop some characteristic features as a power law distributed frequency for the degree. It is worth noting that, power law distributions also appear in other quantities like clustering, betweenness, and average degree of the neighbours.

By the way, scale-invariance is not the only feature of real networks. Rather, by using the framework of graph theory we have a complete series of notions and quantities to describe the different systems. All of them are composed of elements (vertices) connected with each other by edges. If in the simplest case one may focus only on the connections, we may have a more detailed description by keeping record of the strength of the various interactions (even if this it is often impossible to measure). This results in a *weighted graph*. Edges, as in the case of streets between parts of a city, may be one-way only. This happens in food webs and in the World Wide Web; these cases are described by a graph with *oriented* edges. In all the above cases we can define local and global quantities. Concepts like the degree (i.e. the number of edges per vertex) or the clustering (the number of neighbours that are also connected each other) are truly *local* quantities depending upon the state of a single vertex and its neighbours. Measures of centrality, like the betweenness, depend upon the state of the whole system and will be defined in the following as *global*. In between these two scales we have the study of communities that range from a few vertices to the entire graph. In principle other notions are also available in graph theory. For our purposes, we restrict this book essentially to the above set. The exposition of these concepts, together with scale-invariance and the model for scale-free networks are all presented in Part I.

Apart from the theory, most of the book is devoted to discussion of real data. The case of the Internet is one of the most striking examples of scale-invariant networks. The structure of the Internet (defined as the set of computers connected by cables) was properly designed only at the very beginning. During its growth the global structure was no longer under central control; this created traffic problems exactly like those of urban transport in modern cities. Since in the present state the statistical properties of the Internet are now stable, they can be characterized mathematically. Whenever considering this system we always find a hierarchical structure made of a few highly connected computers (hubs) and many poorly connected ones. Similarly in the World Wide Web we have a system of HTML documents (web pages) connected through hyperlinks. In this huge structure (billions of pages), we find again a few hubs (in the sense of pages with millions of links to them, such as Yahoo, Microsoft, Red Hat) and many poorly cited documents. By understanding the graph properties of these structures it is possible to produce reliable search engines, and possibly increase the quality of connections, reducing their cost. In a completely different area, we can spot a similar situation. In every individual cell the genetic code (DNA) has the information on the kind of proteins to assemble in order to accomplish the nearly infinite tasks required to survive. We shall see in the following that proteins interact with each other (even if spotting such interaction is a very delicate and difficult procedure). Surprisingly enough we find again an aristocratic collection of a few hubs interacting with nearly all the others and many poorly connected proteins with one or few interactions.

It would therefore be very tempting to assume that these structures are very similar and all the networks are the same. Actually, only a few characteristics obey such universality. Other quantities differ from each other from case to case. In this book we will focus on both similarities and dissimilarities, trying to convince the audience that while a common reason for the onset of power laws exist, this reason can produce different outcomes.

The structure of the book is rather simple: it starts with some basic notions of graph theory. After that we spend some time in explaining (mainly by use of fractals) why scale-free behaviour is so interesting. Finally we have a part devoted to the models of networks. After the theory we present an overview of the application of the above concepts. The areas selected are those of natural sciences (protein interactions, metabolic and gene regulatory networks, food webs, taxonomies and river networks), information technology (Internet and WWW), and socio-economic sciences (collaboration, cognitive networks, and financial systems). Schematically this is the composition:

- **DEFINITIONS AND METHODOLOGY**
    - **Chapter 1**: a very basic introduction to the concepts needed in graph theory;
    - **Chapter 2**: formation of the graph structures as the clusters and the communities;
    - **Chapter 3**: the concept of scale-invariance and its importance in networks;
    - **Chapter 4**: the origin of scale-invariance. These concepts introduce the last part of methodologies;
    - **Chapter 5**: an overview of the models used to reproduce the real scale-free networks.

- **EXAMPLES**
    - Natural networks
        - **Chapter 6**: scale-free networks in the cell; the case of protein-protein interaction, metabolic, and gene regulatory networks;
        - **Chapter 7**: networks with some scale-free properties in geomorphology; the case of river networks;
        - **Chapter 8**: scale-free networks in ecology; food webs and taxonomic tree of plants and animals;
    - Technological networks
        - **Chapter 9**: Internet, WWW, and e-mail network;
    - Socio-economic networks
        - **Chapter 10**: social and cognitive networks;
        - **Chapter 11**: financial networks;

- **APPENDICES**

PART I

DEFINITIONS AND METHODOLOGY

1. Introduction to graphs

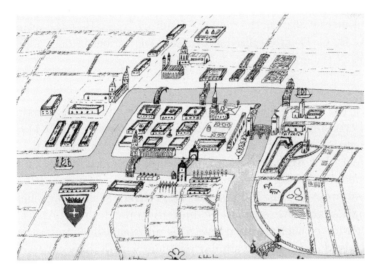

Fig. 1.1 An artistic view of the city of Königsberg (thanks to G. De Masi).

His name was Leonard Euler and he was just a country boy who had never seen Königsberg before. Still the power of mathematics helped a young (he was 29) Swiss mathematician to solve a puzzle about one of the most elegant cities of the eighteenth century. According to the current view, modern graph theory traces back to the mathematician Leonard Euler[2] who was the first scientist to introduce the notion of graphs. The following anecdote eventually resulted in the creation of a new branch of mathematics. Euler wanted to answer a popular question of his time. If we are in the centre of the city of

[2] One of the greatest mathematicians of all times. Born 15 April 1707 in Basel (Switzerland), he died 18 September 1783 in St Petersburg (Russia). Here we discuss his publication, *Solutio problematis ad geometriam situs pertinentis* (Euler, 1736).

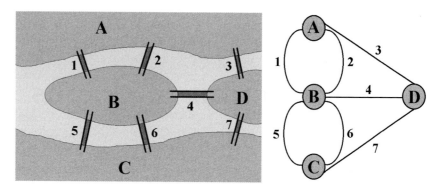

Fig. 1.2 On the left a very schematic map of the town centre of Königsberg at the time of Euler (1736). On the right the resulting graph. Here the various parts of the city (**A, B, C, D**) are stylized as vertices. The various bridges are represented by the edges of the graph.

Königsberg (at the time a Prussian city, now Kaliningrad, Russia) can we do a stroll crossing each of the seven bridges (shown in Fig. 1.1) only once?

The situation was very similar to the simplified map in Fig. 1.2, apart from the fact that we have not reproduced streets, buildings, or the actual shape of the river borders and islands. A brute force solution of this problem could be summarized as follows: we start from a side, we check all the possible paths, and we stop if we find the one desired. Apart from the lack of elegance of such a procedure, this approach does not provide a real solution. Indeed, if we have a similar problem with a different number of 5B bridges we simply repeat the 'exact enumeration' from scratch. Even worse, if at a certain point the number of bridges is greater than seven, the possible paths become so many that it is simply impossible to proceed like this.

A general solution of the problem requires the abstraction of mathematics. The crucial step made by Euler was to encapsulate all the relevant information in one map of Königsberg even more simplified than the previous one. This map is shown on the right of Fig. 1.2.

Here real distances do not matter any more. Different parts of the city (large or small) are described by points called **vertices**. If they are linked (by a bridge) we draw a line (called an **edge**) between them. The map of the city becomes a 'graph'.[3] Through this formalism, the original problem now translates into the more abstract request: 'Is it possible to find a path that passes through all the edges exactly once?'

---

[3] It is worth noting that the same trick is used today to draw maps of the various stations in the undergrounds of modern cities.

Now we have to consider whether the problem has become easier due to this formulation. The answer is both yes and no. No, because 'essentially' (and hopefully) the question has remained the same; yes, because now we have restricted all the attention to these new things we called vertices. All the parts of the city are drawn in the same way (as a point). All of them are equal. Therefore a solution (if it exists) must refer to some intrinsic properties of such objects.

An immediate intrinsic quantity here is the number of edges per vertex, hereafter indicated as the **degree** of the vertex. It is an integer and therefore it is either even or odd. If it is even (consider 2 for now) we realize that the vertex is a crossing point: we can enter the vertex by one bridge and exit by the other. If the number of edges is even but larger than two (i.e. $4, 6, 8$, etc.), the same argument holds. To check that, just divide the edges into pairs. The degree is a number $m = 2n$, where n is the number of pairs of edges. For every couple, we get in through one edge and out through the other. The vertex is visited two, three, or in general n times. Conversely, vertices with odd degree can only be starting or ending points of the path.

That is the solution of the problem! Our request to pass over every bridge exactly once can be satisfied only if the vertices with odd degree are zero (starting and ending point coincide) or two (starting and ending point do not coincide). If you think a little bit and work it out on a piece of paper, you realize that you cannot have a graph with only one vertex with odd degree, so the above ones are the only two possibilities (we see later that the sum of the degrees of the various vertices is an even number, precisely it is twice the total number of edges). Now we go back to the graph in Fig. 1.2 and we discover that none of the above conditions are verified. Actually, all the four vertices have an odd degree. Therefore the path is not possible.

Starting from such problems, graph theory became more and more elaborate. Since the time of Euler, many mathematicians have made important contributions to it. We do not want to provide a formal course in graph theory. Rather, we will focus here only on the basic notions allowing us to study and describe scale-invariant networks. For those who would like to start a detailed study on this topic we can suggest (amongst the many resources available also in electronic form) some introductory books (Bollobás, 1979, 1985; Diestel 2005; West 2001).

> In this chapter we define a graph. We also introduce a way to represent graphs through matrices of numbers. This representation will make some computations particularly easy. When graphs are of very large order (when they have many vertices) the only way to describe them is by means of statistics. Here we will provide some mathematical instruments. Finally, we present the probability distributions we use in the rest of the book.

1.1   Graphs, directed graphs, and weighted graphs

The various networks present in this book are different realizations of the same mathematical object known as a **graph**.[4]

In this chapter we start by presenting basic definitions and then proceed from local to more global quantities. A more traditional explanation of the basic concepts is presented in Appendix B. Here we use only some of these series of notions and I have presented only the quantities necessary for the purpose of this book.

Graphs are assigned by giving a set of the vertices and a set of connections between them. **Vertices** and **edges** are technical terms used in graph theory and we prefer to use them, even if some authors also use site, node, or link (whenever talking about graph theory we will use only the traditional notation for the sake of clarity).

Edges may have arrows, that is they can be crossed in one direction only (we see that this is the case of hyperlinks in an HTML document). In this case the graph is a **directed graph** (a variation of the above concept is that of oriented graph. In this one only one direction is allowed, while in directed both are possible). A further generalization is also possible: one can imagine that a value is assigned to every edge. In the case of transportation networks (a system of pipelines or the Internet cables) this could represent, for example, the maximum load allowed. Whenever this extra information

---

[4] Even if the term 'network' exists in graph theory for a specific concept we refer here to networks as any real system that can be described by means of a mathematical object called a graph. Even if not very strictly, in this book we use the term of graph whenever talking about mathematical properties. Network (web) is instead used for the real systems.

is provided we deal with a **weighted graph**.[5] Further generalizations are possible and used, but for our purposes these categories are sufficient.

The mathematical symbol to indicate a graph composed by n vertices and m edges is usually $G(n, m)$. These parameters n and m are not independent of each other.[6] If we assume that there is only one edge between two vertices there is a maximum number of edges we can draw. Consider that each vertex can establish an edge only with $(n - 1)$ different vertices (and not with itself). This holds for every one of the n vertices. This gives a total number of $n(n - 1)$ possibilities counting every edge twice. The maximum number of edges is exactly one half of that, $m_{max} = n(n - 1)/2$.

If the starting and ending vertices make a difference (as in the case of a directed graph) then we do not have to divide the above quantity by two. In this case, the maximum number of edges is given by $n(n - 1)$.

1.1.1   Adjacency matrix

The structure of the graph $G(n, m)$ can be represented by means of a matrix. Matrices are simply tables of numbers. These tables are very useful for solving problems in linear algebra. We assume that the reader has a basic knowledge of them. As a reminder, we list some of their properties in Appendix D. If needed, a good textbook should be consulted (Golub and Van Loan, 1989).

In the case of graphs we introduce the **adjacency matrix** $\mathbf{A}(n, n)$ whose entries a_{ij} are 0 if vertices i, j are not connected and 1 otherwise. This is a somewhat extended nomenclature. Instead of listing only the edges actually drawn, we decide to write down n^2 numbers, which is more than twice the maximum number of edges ($m_{max} = n(n - 1)/2$) we can draw in a 'simple' (i.e. undirected) graph. The reason for this choice is that through this extended representation it is possible to derive analytically some results of a certain importance. Fig. 1.3 gives an example of the way in which adjacency matrices are defined.

For the moment let us proceed with this representation and consider the form of this matrix. The diagonal elements represent the presence of an edge between a vertex and itself (technically a 'loop'). Unless specified otherwise, we consider those entries equal to 0 (no loops). If loops are allowed (diagonal elements different from 0) then we have n more allowed edges. Therefore the maximum number of edges becomes $m_{max} = n(n - 1)/2 + n = n(n + 1)/2$.

---

[5] Almost everywhere in this book the weight of an edge will be a real positive number.
[6] In a multigraph we can have more than one edge between two vertices, and things are different.

$$A = \begin{pmatrix} 0 & 1 & 0 & 1 \\ 1 & 0 & 0 & 1 \\ 0 & 0 & 0 & 1 \\ 1 & 1 & 1 & 0 \end{pmatrix} \qquad\qquad A = \begin{pmatrix} 0 & 0 & 0 & 1 \\ 1 & 0 & 0 & 0 \\ 0 & 0 & 0 & 0 \\ 0 & 1 & 1 & 0 \end{pmatrix}$$

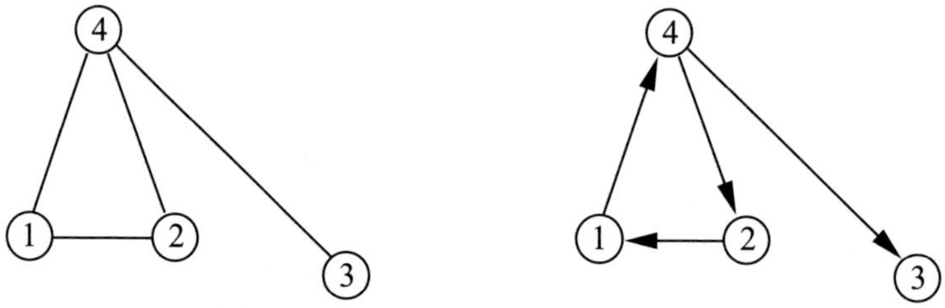

Fig. 1.3 Two simple graphs and their adjacency matrices. Note that for a directed graph (right) the matrix is not symmetric (i.e. it changes if we swap the rows with columns).

Note that this matrix is symmetric (meaning $a_{ij} = a_{ji}$) only in the case of undirected graphs. For directed graphs the elements a_{ij} are generally different from the elements a_{ji} (those symmetric with respect to the diagonal).[7] For example, in the case of only one edge going from vertex 2 to vertex 3 we have that $a_{23} = 1$ and $a_{32} = 0$.

Through this representation we can easily describe the case of weighted graphs. Now instead of giving only 1 and 0, we assign a real number (the weight) to the entry a_{ij}. We obtain an adjacency matrix composed of real numbers for the edges present and 0 otherwise. In the following, we refer to this matrix with the symbol $\mathbf{A^w}(n, n)$. Its elements will then be indicated by a_{ij}^w.

1.1.2   Basic quantities

As above, readers looking for a complete list of concepts should refer to Appendix B. Here we provide definitions for undirected and non-weighted graphs. After this 'simple' case, we also present the other cases.

---

[7] We follow here the overwhelming convention of writing the directed edges of vertex i on the ith row of $\mathbf{A}$. Actually (as pointed out to me by V.D.P. Servedio), it would be simpler to write the edges along the ith column. We see that when the adjacency matrix is transformed in a transition matrix (i.e. every entry is divided by the degree of the vertex) we want $\mathbf{A^T A} = \mathbf{I}$. This result can only be obtained by writing the entries along the columns.

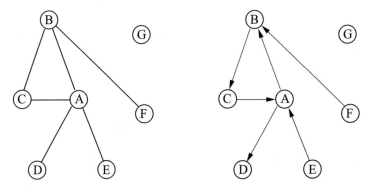

Fig. 1.4 On the left: Graph $G(7,6)$. The **order** of the graph is 7; the **size** is 6. The **degree** of vertex A is 4, the **degree** of vertex C is 2. On the right: A directed graph of the same size. In this case the **in-degree** of vertex A is 2, the **in-degree** of vertex C is 1 and its **out-degree** of vertex C is 1.

- The graph **order** is the number n of its vertices (see Fig. 1.4).
- The graph **size** is the number m of its edges.[8]

  ☕ We recall that in a graph of order n with no self-edges (loops) we can draw a maximum number of edges given by $m_{max} = \frac{n(n-1)}{2}$. This formula is easy to understand. We have n possible starting points and for every one of them $n-1$ destinations. Repeating this procedure we count twice the path from one vertex to another. That is why we must divide by two. This factor in the denominator disappears in a directed graph where we care about the difference between origin and destination of the edges. If we allow loops, we have n more edges and therefore $m_{max} = \frac{n(n-1)}{2} + n = \frac{n(n+1)}{2}$.

  Two immediate limits are present. If no edge is drawn then the graph is **empty** and it is indicated by E_n. If all the edges are drawn, the graph is **complete** and it is indicated by K_n.

- The vertex **degree** is the number of its edges. As mentioned before, the sum of all the degrees in the graph is twice the number of the edges in the graph. This happens because any edge contributes to the degree of the vertex origin and to the degree of vertex destination.

---

[8] This nomenclature is particularly counterintuitive for a physicist. I find it difficult to think that the size of a network is not the number of vertices. Anyway, that is the standard graph theory terminology and I think it is right to use it. In the following, to avoid confusion when we use these terms, we also repeat their definition.

✦ A compact way to compute the degree consists in running on the different columns of a fixed row in the adjacency matrix $\mathbf{A}(n,n)$ looking for all the 1s present. This means that the degree k_i of a vertex i can be computed as

$$k_i = \sum_{j=1,n} a_{ij}. \tag{1.1}$$

In directed graphs this quantity splits in *in-degree* and *out-degree* for edges pointing in and out respectively.

✦ Since in the adjacency matrix the a_{ij} are different from the a_{ji} we have that $a_{ij} = 1$ if and only if an edge goes from i to j. This means that

$$k_i^{in} = \sum_{j=1,n} a_{ji} \tag{1.2}$$

$$k_i^{out} = \sum_{j=1,n} a_{ij} \tag{1.3}$$

In the above definitions all the edges count the same. This is not the case for weighted graphs. In this case, an extension of the degree is made by counting the weights of the edges rather than their number.

✦ The weighted degree k_i^w of a vertex i is then defined as

$$k_i^w = \sum_{j=1,n} a_{ij}^w. \tag{1.4}$$

Some authors (Yook *et al.*, 2001) use the term 'strength' to indicate this quantity. As already noticed (Garlaschelli *et al.*, 2005), Barrat *et al.* (2004*a*) recover that the weighted degree is empirically related to the 'topological' degree k_i by means of a simple relation

$$k_i^w \propto (k_i)^\zeta \tag{1.5}$$

Note that the weighted degree returns the usual degree if matrix $\mathbf{A}^\mathbf{W}$ is replaced by $\mathbf{A}$.

• In an undirected graph two vertices i and j are **connected** if there is a path from i to j. A graph is connected if there is a path for every pair of vertices. A connected component is the maximal subgraph connected in a graph. A directed graph can be either strongly connected if directed paths exist for every pair of vertices, or weakly if paths exist only when considering the edges as undirected.

- The **distance** d_{ij} between two vertices i, j is *the shortest number of edges one needs to travel to get from i to j* (if they are connected). Therefore the neighbours of a vertex are all the vertices that are connected to that vertex by a single edge.

☕ Using the adjacency matrix this can be written as

$$d_{ij} = min\{ \sum_{k,l \in \mathcal{P}_{ij}} a_{kl} \} \tag{1.6}$$

where $\mathcal{P}_{ij}$ is a path connecting vertex i and vertex j.

Note that formally both the sum of the a_{ij} and that of the inverse produce the same result. This is because all the existing edges have a value of 1 and we could also write

$$d_{ij} = min\{ \sum_{k,l \in \mathcal{P}_{ij}, a_{kl} \neq 0} \frac{1}{a_{kl}} \} \tag{1.7}$$

For the case of 'simple' graphs either definition makes equal sense. In the following we see that when considering weighted graphs according to the 'physical' meaning of the weight one or other quantity has a different sense (and of course gives different results). If the graph is directed one has to follow the direction of the edges. Therefore the distances are generally larger than in homologous undirected graphs. In the case of weighted graphs, instead of summing for every step a distance of 1 we can assume that the distance is related to the values of the weight. If the graph represents a distribution network such as a pipeline of water, the weight can represent the cross-section of pipe. Intuitively then one can think that two vertices related by a 'strong' edge (i.e. an edge whose weight is large) are nearer than two related by a weak edge. In this case (e.g. the Internet where weight reports for example the load) we define the distance as the sum of the inverse of the weights.

$$d_{ij}^w = min\{ \sum_{k,l \in \mathcal{P}_{ij}} \frac{1}{a_{kl}^w} \} \tag{1.8}$$

On the other hand, if the network is an electric grid and the weight gives the resistance opposed by current flow on any edge, then the larger the resistance, the larger the distance between two vertices. In this case, the above formula takes a different form.

$$d_{ij}^w = min\{ \sum_{k,l \in \mathcal{P}_{ij}} a_{kl}^w \} \tag{1.9}$$

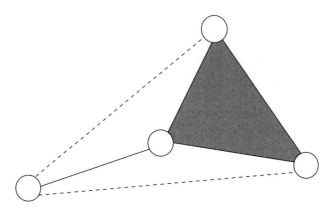

Fig. 1.5 The clustering coefficient of the central vertex is 1/3. This is because its degree is three and its neighbours can be connected each other in three different ways. Of these possibilities (dashed line) only one is actually realized (solid line) and therefore $C_i = 1/3$. The three connected vertices form the coloured triangle. For that reason, sometimes the clustering coefficient of a vertex is defined through the number of triangles it belongs to.

- The **diameter** D of a graph (in this book) is defined as *the largest distance you can find between two vertices in the graph*. Some other definitions (e.g. the average distance) are possible.

- The **clustering coefficient** C_i of vertex i is a measure of the number of edges 'around' the vertex i. C_i is given by the average fraction of pairs of neighbours (of the same vertex) that are also neighbours of each other. A simple example of that is shown in Fig. 1.5. In this case the central vertex i has three neighbours. These can be connected in three different ways, since only one is actually present, this gives $C_i = 1/3$. For the empty graph E_n we have $C_i = 0$ for every i. The maximum value of $C_i = 1$ for every vertex i is obtained for the complete graph K_n. In general we may write the clustering coefficient as the fraction of actual edges over the possible ones between the vertices i, j, k.

Using the formalism of the adjacency matrix we have that for a vertex i whose degree is larger than one

$$C_i = \frac{1}{(k_i)(k_i - 1)/2} \sum_{j,k} a_{ij} a_{ik} a_{jk}. \tag{1.10}$$

The average clustering coefficient C is defined as the average of the C_i over the various vertices i in the graph.

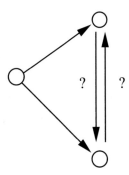

Fig. 1.6 When dealing with directed graphs it is not clear how to close the triangle that forms the basis of the clustering coefficient definition. The same problem holds for weighted networks.

If the graph is directed, the generalization is not straightforward. It is more or less natural to consider the extension of the clustering coefficient only for the in- or out-degree, splitting therefore the contribution one has for the undirected graph. The problem is then to consider which direction of the edge between the neighbours has to be counted (see Fig. 1.6).

 In general one tends to join the two possible directions such that the clustering coefficient takes the form

$$C_i^{in} = \frac{1}{(k_i^{in})(k_i^{in}-1)/2} \sum_{j,k} a_{ji} a_{ki} \frac{(a_{jk}+a_{kj})}{2}, \qquad (1.11)$$

$$C_i^{out} = \frac{1}{(k_i^{out})(k_i^{out}-1)/2} \sum_{j,k} a_{ij} a_{ik} \frac{(a_{jk}+a_{kj})}{2}. \qquad (1.12)$$

For weighted graphs the situation is even more complicated. It is easy to generalize the numerator of the expression above, but we do not know an expression for the denominator. The point is that in a non-weighted graph we can always imagine being able to draw another edge if the graph is not complete. In this case the relative entry in the adjacency matrix will be invariably one. This accounts for the term $(k_i(k_i-1))/2$ giving the total number of edges one can draw. Here instead we also have to assign a weight on the missing edge! Therefore the concept of total weight of triangles is ill-defined.

☢ Different choices are available in order to overcome such a problem. Here we present this definition

$$C_i^w = \frac{1}{\langle a^w \rangle^3 (k_i)(k_i - 1)/2} \sum_{j,k} a_{ij}^w a_{ik}^w a_{jk}^w \qquad (1.13)$$

where the quantity $\langle a^w \rangle = \frac{1}{n} \sum_{ij} a_{ij}^w$ is the average weight of an edge in the graph.

Other choices are possible especially in some real situation where fluctuations from average play a crucial role (i.e. the average is not a representative measure of the set). One possible definition is the following (Barrat *et al.*, 2004*a*, 2004*b*):

$$C_i^w = \frac{1}{(k_i^w)(k_i - 1)} \sum_{j,k} \frac{a_{ij}^w + a_{ik}^w}{2} \theta(a_{ij}^w) \theta(a_{ik}^w) \theta(a_{jk}^w) \qquad (1.14)$$

where $\theta(x)$ is the step function equal to 1 when the argument is larger than 0.

In general, according to the particular case one or another definition can make more sense (Szabó, Alava, and Kertész, 2004).

One important quantity is given by *cliques*. In a undirected graph a $k-clique$ is a complete subgraph K_k such that for any couple of k vertices in the clique, there is one edge connecting them. A **bipartite clique** consists of two sets of vertices such that every vertex of one set is connected with every vertex of the other set. The same concept can be generalized for **tripartite cliques** when we have three sets. More generally cliques can be composed by a number n of sets. This quantity is usually very difficult to measure and visualize. Empirically the number of bipartite cliques $bc(m, n)$ for two sets of order m, n decays very fast (exponentially) for large m, n in almost any network of interest. Simpler methods to describe such systems quantitatively have been introduced to compute the presence of communities.

1.2   Trees

There is one general case in which the networks have a particular characteristic shape. In the case of a distribution network (for example, water supply, but in principle anything), the commodity is delivered to all clients while trying to avoid visiting the same vertex twice. The class of graphs for which this holds are called **trees**. Apart from the formal definition which we can find in Appendix B, we can be more precise by defining

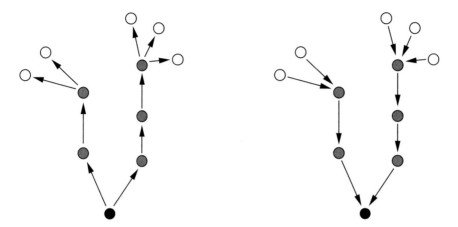

Fig. 1.7 Two examples of special vertices in a tree. On the left (as in a real tree) nutrients flow from the root (dark) to reach the leaves (light). Root and leaves are defined through their in-degree. On the right the case of river networks. Here light-coloured vertices represent the highest zone in the basin (no points uphill). The dark vertex is the outlet of the whole basin. Root and leaves are defined through the out-degree.

- a **cycle** as *a closed path that visits each vertex only once (apart from the end-vertices that coincide).*

From that quantity we have now that

- *a set of vertices connected to each other without cycles is a* **tree**;
- *a set of disconnected trees is a* **forest**.
- For directed trees *the vertices with (out-)in-degree equal to one* (the peripheral vertices of the tree) are called **leaves**. Sometimes it is useful to define a special vertex that is called the **root**. In the case of river networks (as shown in Chapter 7) the root is the vertex (always present) for which the out-degree is zero. By contrast, in food webs people prefer to define as root the vertex whose in-degree is zero (see Fig. 1.7).

On a tree we can still use the quantities defined for graphs. Vertices still have a degree and we can also measure the distances. In a undirected tree there is always a path between any pair of vertices. For directed trees, it is possible that some of the vertices are isolated from the others because the direction of the edges does not allow to join them. Therefore distances are generally larger in directed graphs. The only exception is the clustering coefficient that is zero by construction. Indeed, in the definition given in eqn 1.10 the clustering coefficient is related to the number of triangles (cycles

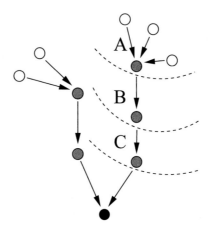

Fig. 1.8 Three different sub-basins nested: A is composed of 4 vertices, B is composed of 5 vertices, C is composed of 6 vertices.

of order 3) present in the graph. Since the tree is an acyclic graph, this quantity is always zero.

Anyway, for this subclass of graphs we can define a new quantity given by the structure of the tree. As shown in Fig. 1.8 a tree can be defined as a set of (sometimes) nested sub-trees. In the picture we show the sub-basins A, B, C whose order (number of vertices) is respectively $4, 5, 6$ (note that in a tree of n vertices, the total number of edges is $n-1$, so whenever talking about trees the size and order of these graphs differ only by one). The size of the various basins is a typical description for a tree. According to the different systems this quantity takes several names. It is called the 'in-degree component' for a general directed tree as well as *drained area* or *basin size* in the language of river networks. In this particular case the vertices correspond to area of the Earth's surface where water is collected from rainfall. The edges represent the directions along which the water flows in the system. For the present system n represents the total amount of points 'uphill' in the tree.

1.2.1   Classification of trees

Because of their simplified form trees often represent a first step in the description of complex objects like graphs. In this book we present a fairly large number of them even if they have a different nature. We try here a little classification and description, according to this nature.

- *Real trees*. The acyclic structure is intrinsic to the physical phenomenon. This is the case of river networks formed by tributary and rivulets of a river. Water follows the downhill slope because of gravity and never climbs back. Therefore it cannot return on the same point. Secondly, water does not divide itself. From a given point a river always follows only one path. This behaviour, by construction, avoid cycles formation.

- *Spanning trees by breadth-first search* As we see in the following this is the case for food webs and that of traceroutes in Internet. Consider that from a vertex in the graph you want to reach rapidly all the other vertices. One way to avoid the possibility to entering a cycle and walking in a circle is given by the breadth-first search (BFS) algorithm. Very schematically, from a starting point (i.e. vertex i) one finds all the first neighbours and writes them in a list; all of them are at a distance one from i. Then for all of these vertices one computes a second list made of their neighbours, provided they are not already in the first list and they are not i. All the vertices in the second list are at distance two from i. Iterating the procedure, one finds different shells of vertices around i, until all the vertices are checked. This algorithm automatically produces a spanning tree (whose root is i) for the graph considered. This kind of structure is similar to the traceroute exploration of the Internet and is often used to determine the skeleton of a transportation network.

- *Trees by classification* The most intuitive example of this phenomenon is given by the natural taxonomy of plants or animals. Starting from field observations generations of naturalists recursively grouped together in larger and larger groups the real species around us. Note that real species such as the laurel (*Laurus nobilis*) or the domestic cat (*Felis catus*), are the only 'real' data. The larger categories into which they are grouped instead are the product of human activity, they cannot be recorded in field observation, they can change if the grouping methodology changes and seldom have an objective meaning.

These two classes of trees are intrinsically different. In one case we have vertices with a similar nature (e.g. the houses to be connected by a water pipeline) in the other the vertices are very different from each other (a vertex corresponding to an actual species and a vertex corresponding to a human classification). In order to define a classification tree, we start by defining as vertices the objects to classify and we connect them according to their correlation.

In the case of botany or zoology, this is very easy. We start from species and we cluster them according to their morphology. Classes of species can be clustered in the same way. Step by step we form a tree composed of different layers. In most cases of interest the distribution of the branches' size is power law distributed (Caldarelli *et al.*, 2004*a*). However, whenever the layers of classification are not well defined, as in the case of financial systems, people usually use another procedure called the **minimal spanning tree**.

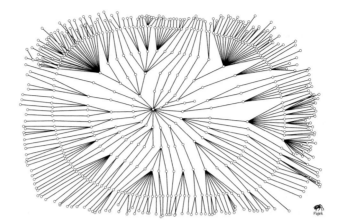

Fig. 1.9 The taxonomy tree of the Coliseum flora in 1815. A plot based on data available in literature (Caneva *et al.*, 2002).

- Define a set of entities that will be represented as vertices in a graph and define an interaction between them that will give the edges (to fix ideas we can use stocks as vertices and price correlation as interactions).
- Using the interaction values we obtain a set of $n \times (n-1)$ correlations of any of the n vertices with all the other $n-1$ vertices. This forms a complete graph with different edges strength. To filter the information, the graph can be transformed in a non-complete one. Two choices are now possible.
  - We can assign a threshold on this weight saying that only the edges with a similarity larger than the threshold are drawn.
  - Otherwise we can classify the different vertices by means of a different procedure. For example, with the minimal spanning tree procedure we can obtain a tree in the following way. Firstly we rank all the similarities between different vertices. Then we draw the first two vertices in the list and we connect them with an edge. We then proceed through the list and draw the second pair of vertices. If by drawing the third pair we close a loop we remove this entry in the list and we proceed along the list. We stop when all the vertices have been drawn, obtaining a picture like that of Fig. 1.9.

Another way in which we can produce trees are the *dendrograms* characteristic of community detection methods. Dendrogram is a word derived from the ancient Greek to indicate a 'tree-like picture', it is another way to represent the correlation between vertices. The difference from the above methods is given by the fact that edges are eliminated one by one from the graph in order to individuate the communities (see Section 2.6.1.2).

1.3   Vertex correlation, assortativity

Following the previous definition for a community, we say that we can spot the presence of a community from vertices with similar properties. How 'similar' two vertices are is usually computed by means of a mathematical quantity called correlation. As usual, since the most immediate properties of a vertex is the degree, we look for the presence of a correlation between vertices with similar degree.

In principle, there is no reason to expect a particular correlation. Actually, in some situations there is a tendency for high-degree vertices to be connected to other high-degree vertices. In this case, the network displays what is called an **assortative mixing** or **assortativity**. The opposite situation, when high-degree vertices attach to low-degree ones, is referred to as **disassortative mixing** or disassortativity (Newman, 2002a).

The correct mathematical way to quantify such a measure is the *conditional probability* $P(k_1|k_2)$ of having a vertex with degree k_1 at one side of the edge given that at the other side of the edge the degree is k_2. In order to compute this quantity we need to define the related concept of joint probability.

1.3.1   Joint vertex probability

For our purposes, we assume that the probability of finding a vertex whose degree is k is given by the limit value of the frequency of the vertices with degree k in the network. That is in the limit of large n

$$P(k) = \frac{n_k}{n} \tag{1.15}$$

where n_k is the number of vertices with degree k and n is the total number of vertices.

It is useful also to define the *joint probability* $P_2(k_1, k_2)$ of extracting an edge and finding that its end-vertices have degrees k_1 and k_2. Note that the joint probability is a different concept from the *conditional probability*. The former gives the probability that in one measurement we find the degrees k_1 and k_2 simultaneously. In the latter, we assume that one degree is known (i.e. k_2) and we would like to know the probability that the other will be k_1.

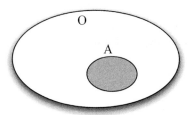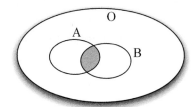

Fig. 1.10 An illustration of the meaning of probability and conditional probability. If O is the space of possible outcomes, we have $P(A) = \frac{\text{area of A}}{\text{area of O}}$. The conditional probability is the value of the probability $P(A \cap B)/P(B)$ i.e. the grey area where both A and B happen, restricted to the space of outcomes B.

These two quantities are related by an axiom of the theory of the probability, that is, the conditional probability of having event A given B is given by

$$P(A|B) = \frac{P_2(A, B)}{P(B)} \tag{1.16}$$

The meaning of this axiom on probability can be easily illustrated. The probability of one event can be thought of as the space occupied by a single event in the area of all the possible outcomes (see Fig. 1.10). This means that the value of probability of one event can be defined as the ratio between the area occupied by a single event and the total area of the possible outcomes. In the case of conditional probability we restrict ourselves to the subspace given by a single event B.

To compute the joint probability, note that if we pick a vertex in the graph at random, the probability that its degree is k is simply $P(k)$. If we instead extract an edge in the graph and follow this edge until we reach one of the vertices, we find a vertex whose degree is k with a probability proportional to $kP(k)$. This is because most of the edges are connected with vertices with large degree and therefore the distribution is biased. In other words, to reach a vertex whose degree is k we need one of them in the graph and this happens with probability $P(k)$; on the other hand we can reach it through its k edges. Therefore this probability that we indicate as $P_{end}(k_1)$ must be proportional to the product $P(k)k$. With some simple statistics it is easy to show that if there is no correlation between the degree of the end-vertices we have an uncorrelated joint probability $P_u(k_1, k_2)$ given by:

$$P_u(k_1, k_2) = \frac{k_1 k_2 P(k_1) P(k_2)}{\langle k \rangle^2}. \tag{1.17}$$

If instead a correlation exists it is not possible to write the form of the joint probability $P(k_1, k_2)$ explicitly in terms of the $P(k)$. Equation 1.16 relating the joint and the conditional probability then takes the form:

$$P(k_1|k_2) = \frac{P_2(k_1, k_2)}{P_{end}(k_2)} = \frac{\langle k \rangle}{k_2 P(k_2)} P(k_1, k_2).$$

(1.18)

A series of useful relations between these quantities is presented below.

☢ We can find the precise form of the $P_{end}(k_1)$ by requiring that the sum over all vertices origin/destination gives 1 (if you extract one edge the end-vertices must have a degree value). This can be done by summing over all the vertices whose degree k_1 is $1, 2, 3, ...$ and is indicated by the symbol $\sum_{k_1}$. If we call this probability $P_{end}(k_1)$ we then find

$$P_{end}(k_1) = \frac{k_1 P(k_1)}{\sum_{k_1} k_1 P(k_1)} = \frac{k_1 P(k_1)}{\langle k \rangle}.$$

(1.19)

We can now immediately compute the value of the joint probability $P_u(k_1, k_2)$ in the case of no correlation between the degrees of the end-vertices. In this case, it is simply given by the product of the two $P_{end}(k_1)$ and $P_{end}(k_2)$, that is

$$P_u(k_1, k_2) = \frac{k_1 P(k_1)}{\langle k \rangle} \frac{k_2 P(k_2)}{\langle k \rangle}.$$

(1.20)

Note that P_u is correctly normalized since $\sum_{k_1, k_2} P_u(k_1, k_2) = 1$.

Otherwise if correlations are present we have in general that the $P_2(k_1, k_2)$ will be proportional to the frequency of observations. This means that we take the number $E_{k_1 k_2}$ of edges between a vertex whose degree is k_1 and a vertex whose degree is k_2 (Pastor-Satorras and Vespignani, 2004).

$$P_2(k_1, k_2) \propto E_{k_1 k_2}.$$

(1.21)

Also in this case the constant of proportionality is computed through normalization

$$P_2(k_1, k_2) = \frac{E_{k_1 k_2}}{\sum_{k_1, k_2} E_{k_1 k_2}}.$$

(1.22)

Now the denominator on the right-hand side is simply twice the total number of edges (we count the same edge twice if the network is undirected) or equivalently $\sum_{k_1,k_2} E_{k_1 k_2} = \langle k \rangle n$. Therefore

$$P_2(k_1, k_2) = \frac{E_{k_1 k_2}}{\langle k \rangle n}. \tag{1.23}$$

From the meaning of $P_2(k_1, k_2)$ we immediately recognize that

$$\sum_{k_2} P_2(k_1, k_2) = P_{end}(k_1) = \frac{k_1 P(k_1)}{\langle k \rangle}. \tag{1.24}$$

That is, if we do not care about the value of k_2 when we extract one edge, we must find the previously computed probability to find an end-vertex whose degree is k_1 (correctly normalized).

Finally, we now use the axiom on conditional probability given by eqn 1.16 and by using eqn 1.19 we obtain the result of eqn 1.18. In other words, to have both k_1 and k_2 at the end of an edge, we firstly have to extract one edge and find k_2 and secondly find the other vertex with degree k_1.

The conditional probability can also be written as

$$P(k_1|k_2) = \frac{P(k_1, k_2)}{P_{end}(k_2)} = \frac{\langle k \rangle P(k_1, k_2)}{k_2 P(k_2)} = \frac{E_{k_1 k_2}}{n_{k_2} k_2} \tag{1.25}$$

where in the last passage we used eqn 1.15 and eqn 1.23

We have two constraints on the conditional probability. The first is given by the normalization condition

$$\sum_{k_1} P(k_1|k_2) = 1. \tag{1.26}$$

The second holds for undirected graphs where the same quantity obeys the detailed balance distribution (Boguñá and Pastor-Satorras, 2002)

$$k_2 P(k_1|k_2) P(k_2) = k_1 P(k_2|k_1) P(k_1) \tag{1.27}$$

This balance equation simply states that the number of edges going from vertices with degree k_1 to vertices with degree k_2 must be equal to the number of edges going from vertices with degree k_2 to vertices with degree k_1.

1.3.2   Average neighbours degree

A less rigorous but more intuitive and simple approach (because the $P(k_1|k_2)$ can be very difficult to measure) is also possible (Pastor-Satorras, Vázquez,

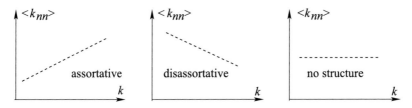

Fig. 1.11 The three possible behaviours of the average degree of the neighbours $\langle k_{nn} \rangle$ versus the degree k of the vertex origin.

and Vespignani, 2001) and actually much more used. We start from a vertex whose degree is k and then compute the **average degree** $\langle k_{nn} \rangle$ **of its neighbours**. This quantity $\langle k_{nn} \rangle$ is in general a function of the degree of the vertex origin and determines the assortativity of the graph.

When plotting $\langle k_{nn} \rangle$ versus k as shown in Fig. 1.11 we can have different behaviour. If the average degree $\langle k_{nn} \rangle$ grows for large k values then big hubs are connected to each other. In this case the network has assortative mixing. On the other hand, if $\langle k_{nn} \rangle$ decreases for large values of k this means that most of the edges of the large hubs are with more or less isolated nodes. In this case the network presents disassortative mixing.

1.3.3   Assortative coefficient

Another measure of assortativity can be obtained through the assortative coefficient r. This number is a particular case of the Pearson correlation coefficient (see 2.6.1.1). In other words, it is another measure of the correlation between the degrees. To compute it we start from the correlation function (Callaway *et al.* 2001; (Newman, 2002*a*))

$$\langle k_1 k_2 \rangle - \langle k_1 \rangle \langle k_2 \rangle = \sum_{k_1, k_2} k_1 k_2 (P(k_1, k_2) - P(k_1) P(k_2)). \qquad (1.28)$$

We now define the assortative coefficient as this correlation function normalized with the variance of the $P(k)$:

$$\sigma^2 = \sum_k k^2 P(k) - \left(\sum_k k P(k) \right)^2. \qquad (1.29)$$

Therefore the assortative coefficient is given by

$$r = \frac{1}{\sigma^2} \sum_{k_1, k_2} k_1 k_2 (P(k_1, k_2) - P(k_1) P(k_2)). \qquad (1.30)$$

In a network with no assortativity we have that

$$P(k_1, k_2) = P(k_1) P(k_2) \qquad (1.31)$$

Network	n	r
Physics co-authorship	52,909	0.363
Biology co-authorship	1,520,251	0.127
Mathematics co-authorship	253,339	0.120
Film actors collaboration	449,913	0.208
Company directors	7,673	0.276
Internet	10,697	−0.189
World Wide Web	269,504	−0.065
Protein interactions	2,115	−0.156
Neural network	307	−0.163
Little Rock Lake	92	−0.276

Table 1.1 Order and assortative coefficient for various networks (Newman, 2002 a).

and the coefficient is zero. Positive values of r signal assortative mixing. Disassortativity corresponds to negative values of r. In Table 1.1 we report some data analysis (Newman, 2002 a). From this data we note that technological and biological networks show disassortative behaviour while social networks are assortative. The reasons for such occurrence are not completely understood.

The same qualitative behaviour can be determined by considering the clustering coefficient of the neighbours of a vertex whose degree is k. Also in this case the correlation between different vertices has the immediate meaning given by the assortativity.

1.3.4   Correlations in weighted graphs

As reported in the previous sections, a weighted graph is given by assigning an individual weight to the various edges. Using the *weighted adjacency matrix* $\mathbf{A^w}$ we have that the strength of edge between vertices i and j is given by the entry a_{ij}^w. As usual, if the graph is not directed the matrix is symmetric. Entry a_{ij}^w equal to 0 means no edge between i and j. In most of the cases, the strength given by the edge weight a_{ij}^w and the degree of the end vertices k_i, k_j are correlated. A large weight (denoting, for example, a large amount of traffic on a Internet cable) is related to the degree of the vertices (computers) connected. To measure such a correlation we study the average[9] $\langle a_{ij}^w \rangle$ versus the product $k_i k_j$. In some cases this quantity behaves

---

[9] In this quantity the average is computed on all the vertices i, j, whose non-weighted degree is k_i, k_j respectively.

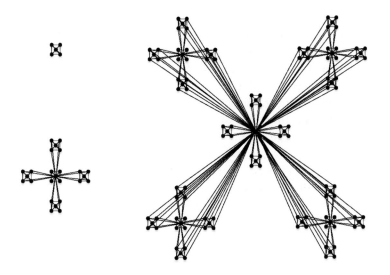

Fig. 1.12 Three steps in the construction of a hierarchical network.

as a power law

$$\langle a_{ij}^w \rangle \propto (k_i k_j)^\theta. \tag{1.32}$$

Whenever we have correlation the value of θ is different from 0. For a transportation network it has been found a value of $\theta = 0.5 \pm 0.1$ (Barrat *et al.*, 2004*a*).

It is interesting to note that the value of the exponent θ is related to the exponent ζ defined in eqn 1.5. Indeed the weighted degree can be thought of as k_i times the average entry in the adjacency matrix so that

$$\begin{cases} k_i^w & \propto & k_i^\zeta \\ k_i^w & \simeq & \langle a_{ij}^w \rangle k_i \propto (k_i k_j)^\theta k_i \end{cases} \tag{1.33}$$

and therefore $\zeta = 1 + \theta$.

1.4   Hierarchical properties of graphs

One of the features of many real networks is their intrinsic modular structure. This aspect makes them even more similar to the fractal structure described in Chapter 3. The key observation is that in many cases we can observe a typical recursive clustering, that is small groups connected to each other to form a larger group and only loosely connected to similar structures on an even larger scale. This property is related to the modularity (defined in Section 2.5) and is a signature of the correlation between the degree of a node and its clustering. The presence of scale-invariance and hierarchical structure gives rise to a particular topology that is captured in a quantitative manner by a suitable scaling law. Consider for example a network with a characteristic degree distribution, one can check what is the average value of the clustering coefficient $C(k)$ for a vertex whose degree is k. In most cases one obtains a scaling law of the kind

$$C(k) \propto k^{-\phi} \tag{1.34}$$

a behaviour that can be reproduced for example by means of hierarchic network[10] built with the procedure outlined in Fig. 1.12. Schematically, starting from a initial module, for example a complete graph K_5, we build a new graph by adding four replicas of the module and connecting the peripheral nodes with the central vertex of the original module. This iteration can be further repeated producing different levels of iteration. This structure has been found in different real networks ranging from actor network to linguistic and biological ones (Ravasz and Barabási, 2003). A discussion of the presence of hierarchy in metabolic networks is presented in Section 6.3.

1.5   The properties of scale-free networks

The aim of this book is to describe the large graphs that you find in nature. Large in some cases means millions of vertices and many more edges. In this situation a convenient description can be obtained only by means of statistical distributions. Some concepts play a crucial role in this book: the first is the statistical distribution $P(k)$ of the degree k; then there is the statistical distribution $P(d)$ of distances d; finally, there is the study of the $C(k)$ representing the clustering coefficient of a node whose degree is k. Note that since the degree k and the distance d are integer numbers we will consider in this book their probability *distributions*, even if for computations we treat these distributions as probability *densities* (see Appendix C).

1.5.1   The distribution $P(k)$ of the degree k

Regardless of the area (biology, computer science, physics, geology, social systems, finance or economics) many structures display a common statistical property (Barabási, 2002; Buchanan, 2002; Dorogovtsev and Mendes, 2003;

---

[10] Actually, also non-hierarchic networks might possess the same property.

Pastor-Satorras and Vespignani, 2004). That is, the *probability distribution for the degree is a power law*. In other words we can write

$$P(k) \propto k^{-\gamma}. \tag{1.35}$$

This sort of 'universality' (even if only qualitative) means that the shape of the degree distribution is similar in the various cases of study. The fact that this shape is a power law has profound implications; we discuss this point thoroughly in Chapter 3. For the moment we only mention that this implies that the system appears the same regardless of the level at which one looks at it. In some cases this could be evidence of a particular and interesting history of the system. In other cases it could simply be the result of statistical noise.

1.5.2   The distribution $P(d)$ of the distances d

In many cases the distribution $P(d)$ of distances d is also similar amongst the various cases of study. In particular the distribution is peaked around small values like $4, 5, 6$. This average value of the vertex vertex distance is supposed to depend logarithmically on the number n of vertices in the network (Szabó, Alava, and Kertész, 2002). This effect is known as the **small-world effect** since in the social graphs where vertices represent individuals, a small number of relationships (edges) can connect two parts of the graph (Milgram, 1967). Like the scale-free properties, the small-world effect has been tested carefully in a variety of different situations. It appears in Internet (Faloutsos, Faloutsos, and Faloutsos, 1999), in the WWW (Broder *et al.*, 2000), in the network of sexual contacts between individuals (Liljeros *et al.*, 2001), as well as in the network of co-authorship where different scientists are connected if they write a paper together (Newman, 2001*a*, 2001*b*).

1.5.3   The correlation between degrees: assortativity and hierarchy

Another typical feature of the scale-free network is the tendency of vertices of a certain degrees to be connected with other vertices with similar or dissimilar degree. As we have already seen in Section 1.3, the first simple way to find this quantity is to select a vertex i and consider the average degree of the neighbours vertices and plot this average as a function of the degree of i.

A similar, but in principle different quantity is given by the plot of the clustering coefficient of a vertex i with respect to its degree k as already shown in Section 1.4. The idea in this case is that if the clustering is high, the

	Network	Order	$\langle k \rangle$	ℓ	C	γ
1	Internet	3700 - 10500	3.6 - 4.1	3.7	0.21 - 0.29	2.1
2	WWW	$2 \cdot 10^8$	15	16	–	2.1
3	Movies	$225,226$	61	3.65	0.79	2.3
4a	Sex (males)	2810	22.63	–	–	1.6
4b	Sex (females)	2810	6.3	–	–	2.1
5	Co-authors	56627	173	4	0.726	1.2
6	Yeast	1870	2.395	–	–	2.4
7	Drosophila	3039	2.40	9.4	–	1.26
8	Stock	240	2.67	18.7	0.08	2.0
9	Words	460902	70.13	2.67	0.437	2.7
10	Flora	282	14	2.65	0.28	2.6

Table 1.2 The general characteristics of several real networks. For each network we indicate the number of nodes, the average degree $\langle k \rangle$, the average path length ℓ, the clustering coefficient C, and the value of the exponent γ for the degree distribution (in the case of WWW we refer to the in-degree one). As a general notice the value of the size of the networks E (i.e. the number of edges) can be found from the definition of average degree. The average degree is by definition twice the number of edges E divided by the number of vertices. This gives $E = N\langle k \rangle /2$ (in the case of directed networks, we should consider the in-degree and out-degree, and in this case the factor two is absent. Here anyway the average degree is the undirected one). The average length ℓ is always taken in the connected component of the graph.

neighbours also, generally have a large degree. Therefore for a disassortative mixing in which a vertex of large degree is connected to many single edge vertices the clustering coefficient and the average degree of neighbours are both decreasing function of the degree.

1.5.4   Data sets

The study of real networks has been based on different data sets available. In Table 1.5.3 we show the main statistical properties of some of them. Data set 1 (Vázquez, Pastor-Satorras, and Vespignani, 2002) refers to the Internet as described by the Autonomous Systems. In data set 2 (Broder et al., 2000) we have $\ell = 6$, when orientation is removed from the edges. When instead the graph is directed the diameter is as large as 500. As for the degree we have a clear power law for the in-degree and an unclear behaviour for the out-degree. Data set 3 (Albert, Jeong, and Barabási, 1999; Barabási and Albert, 1999) has been collected from the Internet Movie Database (http://www.imdb.com). Data set 4 (Liljeros et al., 2001) probably presents a bias in the distribution shape, so that the values of γ as well as the average number of partners is very different (statistically it must be the same when the size of sample goes to infinite). Figures on the number of partners are computed from the plot of the degree distributions. Amongst the possible and various co-authorship networks, we report in data set 5 a collection of papers from high energy physics (Newman, 2001a, 2001b). Data set 6 refers to one of the many analyses (Jeong et $al.$, 2000; Jeong et $al.$, 2001) made on the protein protein interaction network of $Saccharomyces$ $cerevisiae$. The fitting function proposed is of the form $P(k) = a(k + k_0)^{-\gamma} e^{-(k+k_0)/k_c}$ with $k_0 \simeq 1$, $k_c \simeq 20$, and $\gamma \simeq 2.4$. As for the data set 7 it refers to protein protein interaction network of the fruit fly $Drosophila$ $melanogaster$ (Giot et $al.$, 2003). Data set 8 was collected on the Italian Stock Exchange (Garlaschelli et $al.$, 2005). The word co-occurrence network (Ferrer i Cancho and Solé, 2001) in data set 9 is one of the many ways to define a word network. Finally data set 10 refers to taxonomic trees obtained from plant collections (Caretta Cartozo et $al.$, 2006).

2. Graph structures: communities

2.1 Introduction

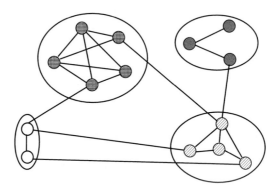

Fig. 2.1 A division in communities in a highly clustered graph.

You don't know about graphs without you have read the previous chapter; but that ain't no matter. All the previous definitions are related to the state of a single vertex (or a few of them, like in the case of two-vertex correlations). In this chapter instead we present quantities that depend upon the connections between a large set of vertices. (In a certain sense, a statistical distribution of *local* properties (such as the degree) is a way to characterize *globally* the shape of a graph. What we mean here is that the definition of the quantities itself is related to many vertices.)

When considering a set of vertices, they might have similar 'properties' (i.e. the same neighbours), so that we can cluster them in the same 'class'. Sometimes a whole series of vertices might be connected to each other with a number of edges larger than in the rest of the graph. In the first case we have what is probably a 'community' in the sense that different vertices have the same 'preference' in their connections. In the second case we have a clustering of vertices. These two concepts are very closely related but in principle different. A particularly clustered subgraph can in almost all cases witness the presence of a community (as shown in Fig. 2.1). On the contrary, communities may not be clustered subgraphs. For example, we can have one set of vertices pointing towards another, in this way these two sets form two communities. Anyway if in the two sets the vertices are not connected to each other (see Fig. 2.9), we do not have any particular clustering. It is clear then that an absolute definition of community is something very difficult to provide. It depends in most cases upon the nature of the vertices considered. Here we make a choice and we use some informal definitions. A *cluster* is a part of the graph where there are more internal edges than external ones.[11] A *community* is a set of vertices sharing the same topological properties. By construction, if a set of vertices has the same edges in common, they are not only a cluster, but also a community since these edges can be considered as a common property. In summary, *a cluster can be always associated with a community of some kind. Communities usually correspond to clustered subgraphs.*

The study of these quantities is not only important in order to characterize the graph topologically. Actually, they can give some information both on the formation of the network and on its functionality. This is because (as is the case for the World Wide Web) in most of the cases we know that the whole network is built by merging different subgraphs. In other cases, the communities select only particular edges among all the possible ones. This helps in determining the traffic or the robustness of the network.

In this chapter we present the state of the art in the methods for computing communities and clustering. For the sake of clarity of exposition we divide such methods into two families: topological analysis and spectral analysis. These methods are made possible by the definition of global quantities depending on the shape of the whole network such as the measures of centrality also presented in this chapter. Readers interested in community detection can find excellent reviews (Danon *et al.* 2007; Newman 2004*b*). devoted to this topic.

---

[11] We call 'cluster' a group of vertices. This definition has nothing to do with the connected component of a graph.

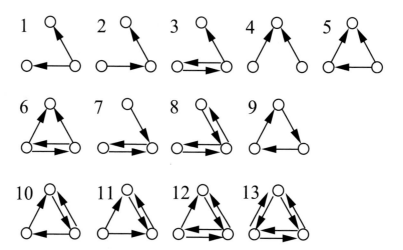

Fig. 2.2 The basic 13 elementary motifs that can be drawn in a directed graph of three vertices. The table made for four vertices motifs has 199 entries.

In this chapter we study some properties of the structure of graphs. We characterize quantitatively the presence of communities and the clustering present in the structures. In this chapter we also use the matrix representation of graphs to compute the community structure. Finally we present some experimental evidence of the statistical properties of scale-free networks.

2.2   Typical subgraphs, motifs

One possible reduction of the complexity of directed graphs can be done by simplifying the graph structure into basic building blocks (Milo *et al.*, 2002). We already know some of the structures shown in Fig. 2.2; for example, the graph number 2 is a path of length 2 while the graph number 9 is a cycle of order three. In general for three vertices it turns out that thirteen possible basic configurations are available (in a directed graph). These configurations called **motifs** help in characterizing the shape of different real networks. In some cases an abundant presence of a particular motif allows one to determine the nature of the whole system. This is because certain recurring motifs can fulfil a particular function in the network.

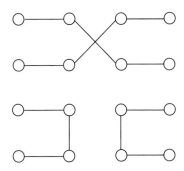

Fig. 2.3 A possible way to rearrange edges keeping the same size, order, and degree sequence.

It is easy to understand that the number of motifs grows very fast with the number of vertices involved. Increasing the number of constituent vertices produces an exponential growth in the number of motifs. For that reason people usually stop very soon and consider only the presence of motifs of order $3, 4$ in the network. Particularly interesting is the fact that in many real networks some motifs are far more frequent than one might expect. Measuring this effect is rather tricky: we start from the graph that we want to analyse, then we build an ensemble of randomized graphs. This is done by considering all the possible graphs with the same order (number of vertices), the same size (number of edges), and the same degree sequence (i.e. any vertex of the graphs in the set has the same degree it had in the original version) of the original one. One way to produce them is to rearrange edges as shown in Fig. 2.3.

At this point we measure the presence of motifs in the original graph and in the 'randomized' set. If the original graph has properties different from that of the ensemble, this is particularly interesting. This seems to happen in many cases (Milo *et al.* 2002; Mangan and Alon 2003), where some motifs are constantly repeated. In particular, their abundance is more than ten standard deviations from the mean expected in randomized graphs. The feed-forward, or filter motif, for example, is common in networks of neurons, but is relatively rare in food webs. A feed-forward motif consists of network vertices x, y and z, in which x has connections to y, and z, and y also has a connection to z. In food webs, where vertices are species, this pattern is carried out only by omnivores (z) that both eat another species (y) and the same food (x) eaten by their prey (in a food web the arrow goes from prey to predator). Instead, in seven different food webs we find abundance of two different motifs. The first is the chain, where one type of prey eats another, which in turn eats another. The second is a diamond-shaped pattern, where one type of prey eats two others, which both eat a fourth type of prey.

Not surprisingly a relative abundance of motifs is also present in the information processing network, where these modular structures can act as logic circuits performing 'and' and 'or' operations. However, in the case of protein interaction networks (Mazurie, Bottani, and Vergassola, 2005) such a relative abundance does not correspond to an effective role. On the other hand, different networks can be effectively characterized by the number and type of cycles present (Bianconi and Capocci, 2003; Bianconi, Caldarelli, and Capocci, 2005). The introduction of weights in the networks considerably modifies the situation (Onnela *et al.*, 2005).

2.3   Classes of vertices

Sometimes the vertices in a graph have a different nature and we can use this information in order for example to compute communities or to determine the spanning trees of the structure. In this book we have two important cases: the World Wide Web and food webs. In the first case the graph is directed, the vertices are the HTML documents and edges can point to or from a certain page. Given the meaning of the graph, it is very likely (and therefore fair to assume) that vertices (i.e. web pages) with a large number of outgoing edges are pages specifically suited (i.e. Yahoo) to reach as many other pages are possible.

On the other hand, if one page has a large number of ingoing edges, it is because its content is probably very important (e.g. an on-line newspaper). The larger the number of edges, the more widely recognized is this importance.

This calls for a new division of vertices, valid only for directed graphs and justified only by the specific character of the case of study (Kleinberg, 1999a; Kleinberg and Lawrence, 2001)

- **hubs** are those web pages that point to a large number of authorities (i.e. they have a large number of outgoing edges).
- **authorities** are those web pages pointed by a large number of hubs (i.e. they have a large number of ingoing edges).

In the case of food webs, the vertices represent species and a directed edge is a predation going from the prey to the predator. According to their habits, the various species may be only prey (basal species), only predators (top species), or both (intermediate species). In this case, vertices whose out-degree is zero are the top species, while those with in-degree zero are the basal species. Sometimes the external environment is also represented as a single species; in this cases all the basal species predate on it and they may have in-degree different from zero. The fact that the edges represent a predation allows another simplification for the graph. All the nutrient originate from environment. For every vertex there must exist a directed path from this 'species' to them. Since in every passage some resources are lost, the shorter paths are more important than the longer ones. Using the BFS algorithm (see Section 1.2.1) from the environment species, one can construct the spanning tree of nutrition flow in a web (Garlaschelli, Caldarelli, and Pietronero, 2003).

2.4   Centrality measures, betweenness, and robustness

We have already seen in Section 1.5.4 that most scale-free networks are characterized by a degree frequency distribution whose tail goes to zero very slowly.

☕ More specifically, in many cases of interest, for large values of the degree k the distribution function $P(k)$ is power law shaped, that is

$$P(k) \propto k^{-\gamma}. \tag{2.36}$$

For the moment we can consider that the above information means that the structure presents few vertices (called 'hubs') with many edges and many vertices with few edges. In some sense, the vertices with the largest degree are the 'most important' in the graph. This concept is particularly clear in the case of the Internet. Whenever hackers want to interrupt the service, they attack the routers with the largest number of connections. Interestingly (and hopefully), some special cases show that this not always true. Therefore the concept of importance (or better *centrality*) can be improved.

Sometimes the situation is rather different. If the graph is the one shown in Fig. 2.4 the hacker should attack the central vertex even if it has a small degree. The notion of importance then depends upon the actual shape of the graph and upon the particular physical meaning of the services provided on the graph. For that reason, different measures of **centrality** of a vertex have been proposed.

These concepts have been used mainly in the field of social science (Freeman, 1977). We describe them here because they can play a role in some scale-free networks. The simplest way to define the most central vertex is to look for the vertex whose average distance from all the others is the minimum one.

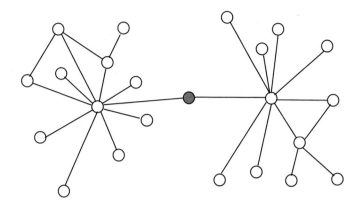

Fig. 2.4 A particular graph where we can disconnect the system acting on the grey vertex with degree 2.

☕ That corresponds to saying that centrality $c(i)$

$$c(i) = \frac{1}{\sum_{j=1,n} d_{ij}}$$

(2.37)

is maximal.

Many graphs reveal a characteristic average distance (this phenomenon is called the small-world effect). This means that this quantity will also show a typical average scale.

Another measure of centrality comes from the dynamical properties of the graph. Whenever the graph is supposed to represent a transportation network (water, electricity, information, etc.), a flux is present on the edges of the graph. It then makes sense to ask how this flux is distributed along the vertices. We start from the simplest choice. We assume that on every edge there is a uniform load. That is to say, any edge has the same capacity of the others. Under this hypothesis, a sensible measure of centrality is given by the number of times that we cross one vertex k in going from one vertex i to another j following the path of minimal length (distance $d(i, j)$). This quantity is called site-**betweenness** $b(i)$ and in formulas is given by

$$b(i) = \sum_{\substack{j,l=1,n \\ i \neq j \neq l}} \frac{\mathcal{D}_{jl}(i)}{\mathcal{D}_{jl}}$$

(2.38)

where $\mathcal{D}_{jl}$ is the total number of different shortest paths (distances) going from j to l and $\mathcal{D}_{jl}(i)$ is the subset of those distances passing through i. The sum runs over all pairs with $i \neq j \neq l$.

It has been conjectured that in scale-free networks (where $P(k) \propto k^{-\gamma}$) the betweenness also is power-law distributed (i.e. $P(b) \propto b^{-\gamma_b}$), characterizing the betweenness frequency distribution $P(b)$. It is possible to connect the properties of the betweenness distribution to that of the degree distribution (Goh, Kahng, and Kim, 2001; Barthélemy, 2004). When the network has a degree distribution whose exponent is $3 \geq \gamma > 2$ the exponent γ_b is near 2 (Goh, Kahng, and Kim, 2001). More particularly it can be shown that

$$\gamma_b \geq \frac{\gamma + 1}{2} \qquad (2.39)$$

where the equality holds if the network is a tree (Barthélemy, 2004). An immediate generalization of this quantity is given by the edge-**betweenness** where now the number of paths considered are those passing for a certain edge.

When the number of vertices increases, one has to use some care in order to compute this quantity along the graph. Simple algorithms tend to increase the computation time very easily, so that in some cases a series of methods to obtain an approximate value of the betweenness has been produced. Very recently a fast and efficient algorithm to compute such quantity exactly has been presented (Brandes, 2001).

A particular interest for applications is to check whether a particular graph can survive to a certain number of deletions of both edges and vertices. Traditionally such deletions are divided into two classes (Albert, Jeong, and Barabási 2000, 2001). The *random* failure where a vertex is randomly removed regardless of its importance (degree) or centrality in the network and the *attack* where a vertex is removed with a probability related to its importance (usually the degree). The key quantity to monitor in order to check if the properties of the network are unaffected is the change in the value of the diameter D or of the average distance with respect to the fraction f of removed vertices.

2.5   Clustering detection, modularity

A set of similar vertices can form a subgraph larger than the three or four vertices usually considered in motifs. Sometimes the edges are sufficiently dense for more than one division into subgraphs to be possible. A quantity called *modularity* has been introduced (Newman and Girvan, 2003) in order to decide if one division is better than another (Newman, 2006). This method evaluates the outcome of a division in clusters; it analyses a particular subgraph and counts how many edges are inside the cluster and how many point outside the cluster. If the number of internal edges is large and the number of edges between different clusters is small then the division is appropriate. In the case of a clique, we can spot the presence of the whole clique without detecting the communities inside.

The starting point is to find a suitable division of the graph into g subgraphs. To detect whether the division is 'good' or not, we define a $g \times g$ matrix $\mathbf{E}$ whose entries e_{ij} give the fraction of edges that in the original graph connect subgraph i to subgraph j. The idea is that *we want the largest possible number of edges within a community and the lowest possible number of edges between different communities.*

The actual fraction of edges in the subgraph i is given by element e_{ii}; the quantity $f_i = \sum_{j=1,g} e_{ij}$ gives the probability that in a random partition an end-vertex of a randomly extracted edge is in subgraph i. In order to define if this partition is good, we must compare the measured e_{ii} with the corresponding quantity in a null case (e.g. a random partition). The probability that an edge begins at a node in i, (i.e. f_i) is multiplied by the fraction of edges that end at a node in i (i.e. again f_i). In this way the expected number of intra-community edges is given by f_i^2.

☕ The **modularity** of a subgraph division is then defined as

$$Q = \sum_i e_{ii} - f_i^2 \tag{2.40}$$

and represents a measure of the validity of a certain partition of the graph.

In the limit case where we have a random series of communities, the edges can be with the same probability in the same subgraph i or between two different subgraphs i, j. In this case $e_{ii} = f_i^2$ and $Q = 0$. If the division into subgraphs is appropriate, then the actual fraction of internal edges e_{ii} is larger than the estimate f_i^2, and the modularity is larger than zero.

Surprisingly, random graphs (which, as we shall see in the following, are graphs obtained by randomly drawing edges between vertices) can present partitions with large modularity (Guimerà, Sales-Pardo, and Amaral, 2004). In random networks of finite size it is possible to find a partition which not only has a non-zero value of modularity but even quite high values. For example, a network of 128 nodes and 1,024 edges has a maximum modularity of 0.208. While on average we expect a null modularity for a random graph, this does not exclude that with a careful choice we can obtain a different result. This suggests that these networks that seem to have no structure actually exhibit community structure due to fluctuations.

2.6   Communities in graphs

Now that we know how to measure the similarity of the vertices we can introduce the methods that can be used to compute actually the presence and size of communities. Two different approaches are possible. The first selects suitable subgraphs on the basis of special vertices or special edges present in the graph. For example, if we spot few vertices with many edges in common, that is a natural signature of a community. We call this method *topological analysis*. The second approach (used mainly by graph theorists and mathematicians) is based on a method called *spectral analysis*. This method selects communities from the properties of the eigenvectors of the adjacency matrix $A(n, n)$ of the graph. In the first case we can subdivide the approach further, using a nomenclature introduced by Newman (2004*b*). In one case we can build communities by recursive grouping of vertices (bottom-up process) and this process is called an *agglomerative method*. The other possibility is to operate a top-down process with a recursive removal of vertices and edges. This latter process is a *divisive method*. In principle, a similar approach could also be done for the spectral analysis, but a distinction between divisive and agglomerative methods in spectral analysis is far less intuitive.

2.6.1   Topological analysis

2.6.1.1   Agglomerative methods   Once the network is given, the *similarity* between vertices i and j can be measured according to several formulas. In almost all cases the similarity concepts can be defined by means of a suitable 'distance' or viceversa. Essentially if two vertices are similar they must be close together. There is no immediate choice between these various methods. Rather the study of the actual real system represented by the graph is the only way to determine which of these quantities is more sensible than others.

- **Structural equivalence** Two vertices have a structural equivalence if they have the same set of neighbours. This strong requirement can be relaxed when defining a distance between the two sets of neighbours. Obviously, the smaller this distance, the greater the equivalence. Different real situations call for different measures of such a quantity. This concept can be used for protein classification (Brun *et al.*, 2003; Vázquez *et al.*, 2003). A somewhat different notion (even if related) is to define two vertices similar if their neighbours are similar (Leicht, Holme, and Newman, 2006). In such a way the measure of similarity arise self-consistently from the adjacency matrix.

The simplest definition of similarity is more or less equivalent to the *Hamming distance*[12] obtained by counting the number n_{ij} of similar neighbours between vertices i and j. In order to have a distance properly defined (i.e. non-negative) we can consider the following quantity (Burt, 1976; Wasserman and Faust, 1994)

$$x_{ij}^H = \sqrt{\sum_{k \neq i,j} (a_{ik} - a_{jk})^2}. \tag{2.41}$$

A similar approach to determining the similarity of two vertices has been used to determine the function of unknown protein in the protein protein interaction network of the baker's yeast *Saccharomyces cerevisiae* (Brun *et al.*, 2003). The idea is to use the set S_i of vertices neighbours to i and the set S_j of vertices that are neighbours to j. Denoting by $N(S_{i,j})$ a function that gives the number of elements (vertices) in sets $S_{i,j}$, we have that the quantity

$$x_{ij}^S = \frac{N(S_i \cup S_j) - N(S_i \cap S_j)}{N(S_i \cup S_j) + N(S_i \cap S_j)} \tag{2.42}$$

has the desired property to be 0 when S_i and S_j coincide, growing instead when the sets differ. An example of the application of this quantity is reported in Fig. 2.5. In the case of the protein protein interaction network, is it possible to classify the 11 per cent of the proteins present, with a prediction success of about 58 per cent.

The above requirement can be put in the more complicated language of the adjacency matrix. The x_{ij}^S is then given by

$$x_{ij}^S = \frac{\sum_k a_{ik} + \sum_k a_{jk} - 2\sum_k a_{ik}a_{jk}}{\sum_k a_{ik} + \sum_k a_{jk}} = \frac{\sum_k a_{ik} + a_{jk} - 2a_{ik}a_{jk}}{\sum_k a_{ik} + a_{jk}} \tag{2.43}$$

A much weaker assumption regards the symmetry of vertices, in the sense that we can measure the correlation of the degree sequences between two vertices, but only if they have the same degree (Holme, 2006).

- **Correlation coefficient.** In this case the distance between vertices is computed by considering the mean and the variance of the values along a row (or column) of the adjacency matrix. We use these quantities in

---

[12] Named after the US mathematician Richard Wesley Hamming (1915 98) known for his work in information theory.

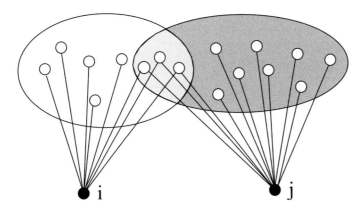

Fig. 2.5 The union of the two sets contains sixteen vertices, the intersection three. The enumerator of x_{ij}^S is given by the vertices belonging only to one or the other. In this case this number is thirteen. The distance x_{ij}^S between the two vertices i, j is then $\frac{16-3}{16+3} = \frac{13}{19} \simeq 0.68$.

order to compute the correlation C_{ij}^P, between vertices i, j. Correlation is different from distance. Now if the two vertices are the same (i.e. $i = j$) the value C_{ij}^P must be 1 while their distance is 0. If i, j are very different they must have a large distance while their correlation is around 0.

 In formulas mean and variance are:

$$\mu_i = \frac{1}{n} \sum_j a_{ij}, \quad \sigma_i^2 = \frac{1}{n} \sum_j (a_{ij} - \mu_i)^2 \tag{2.44}$$

and the correlation coefficient[13] is defined as

$$C_{ij}^P = \frac{\frac{1}{n} \sum_k (a_{ik} - \mu_i)(a_{jk} - \mu_j)}{\sigma_i \sigma_j} \tag{2.45}$$

which is equal to 1 when $i = j$ as expected.

Note that we always use the entries of the unweighted adjacency matrix $A(n, n)$ so that the quantity μ_i corresponds to the normalized degree k_i/n of the vertex i. When the matrix is directed μ_i gives in-/out-degree according to whether the sum is over the rows or the columns.

---

[13] This coefficient is often named after Karl Pearson an English mathematician who was born in London in 1857 and died in Coldharbour, Surrey, England in 1936. Founder of the journal *Biometrika*, he worked mainly in the statistical studies of populations.

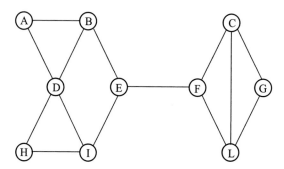

Fig. 2.6 A simple graph, on which we apply the iterative division of the method of Girvan and Newman. The result is shown in the next figure.

2.6.1.2   Divisive methods: the algorithm of Girvan and Newman

Instead of grouping together vertices, we can decide to remove some vertices from the initial graph (this can be done with edges as well). If this procedure is based on suitably chosen measures of centrality we end with subsets that are the corresponding communities we are looking for (Girvan and Newman, 2002). In practice we select the edge of largest betweenness in the graph and then we remove it. Edges have large betweenness if they connect parts of the graph that would be separate otherwise. This seems a rather good heuristic method for detecting separate subsets, since it selects and removes the bottleneck edges separating different communities.

Note that this method works reasonably well for sparse graphs (this is the case in most situations). In this case, we can discover the community structure simply by removing the first edges. On the other hand, when dealing with dense graphs, this procedure can be a little tricky. This idea is the prototype of a class of edge-removing methods.

- **Edge-betweenness.** The first and original approach (Girvan and Newman, 2002) is based on **edge-betweenness**. This quantity is computed on all the edges of the graph. Recursively the edge with the largest betweenness is removed until no edge remains. During this process the graph becomes disconnected. This is the moment when the structure of communities emerges. We write all the vertices in one subgraph on one side. On the other side we put all the other vertices.

    To fix the ideas let us consider the simple case in Fig. 2.6. We have ten vertices and the series of edge removals in this case is the following. The first one is the edge **E-F**: this corresponds also to a separation of the vertices in two distinct graphs. Whenever this happens we write the edge in bold style. As a consequence of this splitting we have that one community is composed of the first six vertices and another of the

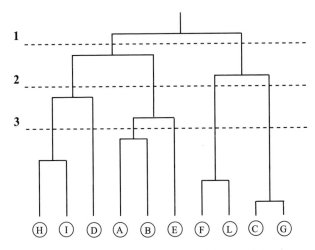

Fig. 2.7 The dendrogram obtained by sequentially removing the edges with the largest betweenness. Line 1 represents the system after the edge E-F is removed, leaving two communities. Line 2 represents the system after the edge G-L is removed, leaving four communities: (A,B,E), (D,H,I), (F,L) (C,G). Finally line 3 is the system after removing the edge B-E. At the end of the process, many edges have the same (largest) value of betweenness. Therefore the edge to remove is selected randomly.

four vertices left. The subsequent series of deletions are the edges E-I, B-D, **A-D**, C-F, C-L, D-H, **G-L**, **D-I**, **B-E**, **A-B**, **H-I**, **F-L**, **C-G**. Often, especially at the end of the process, many edges have the same (largest) value of betweenness. If more than one edge shares the same value of betweenness the one to be removed is selected at random. All this series of divisions can be represented by a dendrogram like the one shown in Fig. 2.7. We start from the whole graph which corresponds to the root of the dendrogram. When we have the first division we bifurcate the dendrogram by placing on one side all the vertices in one of the communities and on the other side the vertices that are left. This recursive splitting ends at the leaves of the tree, which represents a situation where all the vertices of the graph are disconnected.

As discussed below, the main inconvenience of this method is that it eventually splits the graph into all the vertices, regardless of the real number of communities present.

- **Approximated edge-betweenness.** Instead of computing the betweenness by considering all the possible paths through an edge, one can approximate this quantity by computing only the paths from a random selection of vertices (Tyler, Wilkinson, and Huberman 2003).

With such an approach, we have a substantial increase in speed as well as a stochastic measure for the communities. The number of times we find a particular vertex in the same community gives a measure of the reliability of such assignment.

- **Loop counting methods**. This time the edges are selected according to the number of small cycles they belong to. The idea is that edges joining communities are bottlenecks and it is unlikely they belong to a cycle. If this were the case another path would be present between the same communities (Radicchi *et al.*, 2004). Therefore by measuring the number of small cycles (in this case triangles or cycles of order 3) a particular edge belongs to, we have a measure of vertex centrality. We may expect that the lower this number is, the greater its importance for joining different communities. The key quantity is now the *edge clustering coefficient*

$$C_{ij} = \frac{z_{ij} + 1}{min(k_i - 1, k_j - 1)} \tag{2.46}$$

where i, j are two vertices of degree k_i, k_j respectively, and z_{ij} are the number of triangles an edge belongs to. The algorithm iteratively deletes the edges with lowest values of C_{ij} and recomputes this quantity for the edges left.

2.6.1.3   Communities from k-cliques.

A slightly different approach consists in the evaluation of complete subgraphs present in the system (Palla *et al.*, 2005). This method is particularly suitable for very dense graphs, even better if they are constituted by complete subgraphs ($k-$cliques in undirected graphs). This is exactly the case for boards of directors or coauthorship and more generally any social network that is obtained by projecting a bipartite graph (e.g. those in Fig. 10.2 or in Fig. 11.2). The definition of community at the heart of the method consists in considering the union of all the $k-$cliques (K_k) that can be reached from each other by a series of adjacent $k-$cliques. In the original paper the search for such structures in a graph is made by first looking at all the complete subgraphs K_k and then studying the overlap between two of them. In the real data analysis the author filtered the information by removing (if present) directionality on the edges and keeping the edges if the connection is larger than a fixed threshold w^*. When analysing data, it is important to test different similar values of k and w^* to check the robustness of the structure found. A way to define the values used in the decomposition of the graph is related to the problem of percolation (see Section 3.5.1.1). Once a value for the k (usually between 3 and 6) has been defined the value of the threshold w^* is tuned in such a way as to find the maximum possible number of communities avoiding a giant one. Authors of the original paper (Palla *et al.*, 2005) suggest of lowering w^* such that the largest community is twice the size of the second one. In any case the fraction f of the edges above the threshold must be more than $1/2$. When the graph is not weighted they select the smallest value of k for which a giant component disappears. One can also produce a coarse-grained version of the graph where the communities take the place of the original vertices and study the statistical distribution for this network of communities. In the case of coauthorship network (cond-mat arXiv), protein interaction network (DIP database for *S. cerevisiae*), and word association network, this structure is characterized by an exponential decay of the kind

$$P(k_{com}) \propto e^{k_{com}/k^0_{com}} \tag{2.47}$$

The distribution of community sizes s_{com} is also a power law

$$P(s_{com}) \propto s_{com}^{-\tau} \tag{2.48}$$

with τ between 1 and 1.6.

2.6.2   Spectral analysis

This is a completely different class of methods where the structure of communities is determined by the eigenvalues and eigenvectors of suitable functions of the adjacency matrix $\mathbf{A}$ (Hall, 1970; Seary and Richards, 1995; Kleinberg, 1999*b*). The functions of $\mathbf{A}$ that are most used are the normal matrix (actually a 'normalized' matrix) and the Laplacian matrix. The basic notion necessary to understand this approach are presented in Appendix D. A more complete exposition can be found in basic texts on linear algebra (e.g. Golub and Van Loan, 1989).

2.6.2.1   Normal matrix   The elements of the normal matrix are those of the adjacency matrix divided for the degree of the node. In linear algebra that can be written as

$$\mathbf{N} = \mathbf{K}^{-1}\mathbf{A} \tag{2.49}$$

where the matrix $\mathbf{K}$ is a diagonal matrix that has on the diagonal element k_{ii} the degree of vertex i.

☕ In general we can write $k_{ij} = \delta_{ij} \sum_{l=1}^{N} a_{il}$, or more clearly $k_{ii} = \sum_{l=1}^{N} a_{il} = k_i$ and $k_{ij} = 0$ when $i \neq j$. We have that

$$n_{ij} = \frac{a_{ij}}{k_i}. \tag{2.50}$$

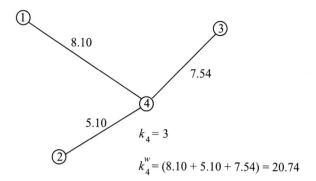

Fig. 2.8 A simple example of a graph on which we compute k, k^w.

In the extended form we have

$$\mathbf{N} = \begin{pmatrix} a_{11}/k_1 & a_{12}/k_1 & \dots & a_{1n}/k_1 \\ a_{21}/k_2 & a_{22}/k_2 & \dots & a_{2n}/k_2 \\ \dots & \dots & \dots & \dots \\ a_{n1}/k_n & -a_{n2}/k_n & \dots & a_{nn}/k_n \end{pmatrix}. \qquad (2.51)$$

Note that the matrix product is not symmetric, and therefore $\mathbf{K}^{-1}\mathbf{A} \neq \mathbf{A}\mathbf{K}^{-1}$. The normal matrix can be defined only for non-isolated vertices (those for which the degree is different from zero).

Consider the particular case of a large graph, where we are on the vertex $i = 4$ whose degree is 3 and this vertex is connected with vertices $1, 2$, and 3, this would correspond to having the ith row (i.e. the fourth row) of the matrix given by

$$1/3, 1/3, 1/3, 0, 0, \dots, 0. \qquad (2.52)$$

These entries of the matrix can be regarded as the *probabilities of passing directly* from node i (i.e. 4) to one of the neighbours (i.e. $1, 2, 3$). In the case of a simple graph, one assumes that any edge counts the same, so that it is equally probable to pass from one vertex to any other of the neighbours. Because of this probabilistic property, this matrix is also known as the *transition matrix*.

 Note that if the graph is directed, the matrix $\mathbf{N}^{\mathbf{T}}$ defined by swapping the rows with the columns in $\mathbf{N}$ can be viewed as the normal matrix of a graph where the direction of the arrows has been exchanged. Coming back to the probabilistic meaning, if $\mathbf{N}$ transfers from one vertex to another, we can come back using $\mathbf{N}^{\mathbf{T}}$. Therefore by recursively applying

N followed by $\mathbf{N}^{\mathbf{T}}$ we must obtain the identity matrix. That is,

$$\mathbf{N}^{\mathbf{T}}\mathbf{N} = \mathbf{I}. \tag{2.53}$$

Since the matrix product is not symmetric, all the above requirements are fulfilled if, *against the standard notation*, entries in the adjacency matrices are written along the columns (see also Section 1.1.1).

In the case of weighted graphs these probabilities can be assumed to be proportional to the weights. Following the above example we can compute the probabilities from the weights. For example, we can say that the weight of edge $(1-4)$ is 8.10, the weight of edge $(2-4)$ is 5.10, and the weight of edge $(3 - 4)$ is 7.54 (see, for example, Fig. 2.8). This would give a generalized degree $k_4^w = 8.10 + 5.10 + 7.54 = 20.64$. The resulting fourth row of the normal matrix

$$\mathbf{N}^{\mathbf{w}} = (\mathbf{K}^{\mathbf{w}})^{-1}\mathbf{A}^{\mathbf{w}} \tag{2.54}$$

whose components can be written as

$$n_{ij}^w = \frac{a_{ij}^w}{k_i^w} \tag{2.55}$$

is represented by the following entries

$$0.39, 0.25, 0.36, 0, 0, \ldots, 0. \tag{2.56}$$

Note that by construction (and also for the probabilistic argument above) the sum of the entries along a row is equal to 1.

2.6.2.2   Laplacian matrix   Another interesting function of the adjacency matrix $\mathbf{A}$ is given by the Laplacian matrix $\mathbf{L}$. In the language of matrix operations, this matrix is given by

$$\mathbf{L} = \mathbf{K} - \mathbf{A}. \tag{2.57}$$

Where $\mathbf{K}$ is the degree matrix previously defined where the elements are all zero apart from the ones the diagonal where they are equal to the degree k_i of the vertex i. This allow to write

$$\mathbf{L} = \begin{pmatrix} k_1 & -a_{12} & \ldots & -a_{1n} \\ -a_{21} & k_2 & \ldots & -a_{2n} \\ \ldots & \ldots & \ldots & \ldots \\ -a_{n1} & -a_{n2} & \ldots & k_n \end{pmatrix}. \tag{2.58}$$

Or for every component

$$l_{ij} = \delta_{ij}k_i - a_{ij}. \tag{2.59}$$

Using our previous little example this means that the $i-th$ row of the matrix $\mathbf{L}$ (for unweighted graph) would be

$$-1, -1, -1, 3, 0, \dots, 0. \tag{2.60}$$

while for the weighted graph, as shown in Fig. 2.8, we have

$$-8.10, -5.10, -7.54, 20.64, 0\dots, 0. \tag{2.61}$$

☞ In the case of weighted networks we substitute k_i and a_{ij} with their respective correspondent k_i^w and a_{ij}^w. This produces the new weighted Laplacian matrix given by

$$\mathbf{L^w} = \mathbf{K^w} - \mathbf{A^w} \tag{2.62}$$

whose components are

$$l_{ij}^w = \delta_{ij} k_i^w - a_{ij}^w. \tag{2.63}$$

The name 'Laplacian' given to this matrix comes from the fact that the i-th row of the matrix gives the value of the Laplacian operator on the vertex i.

♣ As explained in Section 4.1.1, the Laplacian operator is defined as

$$\nabla^2 \phi(x, y, z) = \frac{\partial^2 \phi(x, y, z)}{\partial x^2} + \frac{\partial^2 \phi(x, y, z)}{\partial y^2} + \frac{\partial^2 \phi(x, y, z)}{\partial z^2} \tag{2.64}$$

That is, a Laplacian operator applied on a certain function (in this case $\phi(x, y, z)$) returns the sum of the three second derivatives with respect to x, y, z. When the function is defined only on discrete values, we use the finite differences instead of derivatives. That is

$$\frac{\partial \phi(x, y, z)}{\partial x} \rightarrow \phi(x+1, y, z) - \phi(x, y, z) = \phi(x, y, z) - \phi(x-1, y, z) \tag{2.65}$$

To compute the second derivative we take first the forward difference and then the backward one. In this case, the Laplacian operator defined in eqn 2.64 takes the form

$$\nabla^2 \phi(x, y, z) = \sum_{\xi, \eta, \zeta} \phi(\xi, \eta, \zeta) - k\phi(x, y, z) \tag{2.66}$$

where ξ, η, ζ indicates the coordinate of one neighbour of x, y, z and k is the total number of such neighbours. By comparing this expression with any one line of the matrix $\mathbf{L}$ we immediately see that the sum of the element along the row gives exactly the Laplacian operator (with a minus sign).

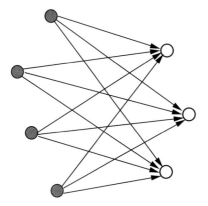

Fig. 2.9 A very simplified example of a characteristic structure of the World Wide Web. The dark vertices are the *hubs* of the graph, the light ones are the *authorities*. Note that neither the light vertices nor the dark ones form a cluster.

2.6.3   Thematic divisions

Whenever it is possible to distinguish the role of various vertices, we can use this information to detect communities, the best-known example is that of the HITS algorithm.

For a directed network, it is possible to extract community information just from the edges structure. This algorithm call HITS has been proposed on an empirical basis (Kleinberg, 1999a, 1999b) in order to find the main structures in the World Wide Web. Web pages are divided in two categories: the **hubs** and the **authorities** (see Fig. 2.9).

Those quantities are defined in a self-organized way by the dynamics of the World Wide Web. By the creation of a hyperlink from page p to q, the author of page p increases the authority of q. So the first recipe in order to define the authority of a WWW site would be to consider its in-degree. This is only an approximation; in the World Wide Web, many edges are created without specific reference to the authority of a page (i.e. the hyperlinks to return to the home page). To partly overcome this problem we need to establish at the same time also the counterpart of the authoritative sites. Those are *hubs*. A hub is defined as a WWW site pointing to many authorities. It can be demonstrated that a specific relation exists between the two quantities.

The algorithm works by starting with a subset of the Web (obtained by means of a text-searching algorithm). On that subgraph we remove the internal edges and the hyperlinks pointing to the webmasters. The iterative procedure starts by assigning to every vertex p (a page) a non-negative

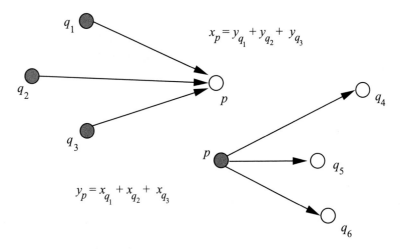

Fig. 2.10 The definition of the hub weight x_p and authority weight y_p.

authority weight x_p and a non-negative hub weight y_p. The idea is to solve simultaneously all the equations in order to find the set of values of x and y satisfying these requirements. This corresponds to writing

$$x_p = \sum_{q \to p} y_q$$

$$y_p = \sum_{p \to q} x_q \tag{2.67}$$

where $q \to p$ runs on all the pages q pointing to p and $p \to q$ runs on all the pages p pointing to q (see Fig. 2.10). From the point of view of linear algebra the solution of such a problem consists in finding the set of eigenvalues and eigenvectors of the matrices $\mathbf{AA}^\mathbf{T}$ and $\mathbf{A}^\mathbf{T}\mathbf{A}$.

☕ This can be shown arranging in a vector $\mathbf{x}$ the various x_p for every vertex p. Similarly, the various y_p can be arranged in a vector $\mathbf{y}$. Then the above eqn 2.67 between $\mathbf{x}$ and $\mathbf{y}$ can be written as

$$\mathbf{x} = \mathbf{A}^\mathbf{T}\mathbf{y} \text{ or } \mathbf{y} = \mathbf{A}\mathbf{x} \tag{2.68}$$

substituting one equation in the other we simply find

$$\mathbf{x} = (\mathbf{A}^\mathbf{T}\mathbf{A})\mathbf{x} \text{ and } \mathbf{y} = (\mathbf{A}\mathbf{A}^\mathbf{T})\mathbf{y} \tag{2.69}$$

Therefore the solution of the conditions required is represented by the eigenvectors of the matrix operator $\mathbf{A}^\mathbf{T}\mathbf{A}$ for $\mathbf{x}$. Similarly the eigenvalues of $\mathbf{AA}^\mathbf{T}$ give the solution for vector $\mathbf{y}$.

Such an algorithm efficiently detects the main communities, even when these are not sharply defined. However, it becomes computationally expensive when one is interested in minor communities, which correspond to smaller eigenvalues. For a description of the numerical solution of this algorithm (and related problems) see Section 9.6.1.

2.6.4 Eigenvectors and communities

Any matrix (see Appendix D for a brief explanation and textbooks (e.g. Golub and Van Loan, 1989) for a general view) is characterized by a set of quantities. These are the eigenvalues λ^j and the eigenvectors $\mathbf{x}^j$ that enter in the equation

$$\mathbf{A}\mathbf{x}^j = \lambda^j \mathbf{x}^j. \tag{2.70}$$

A matrix of size n (n rows and n columns) has n eigenvalues and n related eigenvectors (i.e. j goes from 1 to n). Some of these n eigenvalues (in general complex numbers) can coincide. In some particular cases (if the matrix is symmetric, if the matrix has a physical meaning (describing probability rate, etc.), the series of eigenvectors has some interesting properties. For example in the case of the normal matrix $\mathbf{N}$ a constant vector is an eigenvector of the matrix with eigenvalue equal to 1. By constant vector we mean a vector whose constant components are all the same. To simplify the notation, let us indicate this eigenvector simply by $\mathbf{x}$. This property derives directly from the fact that the sum of the elements along a row in the matrix $\mathbf{N}$ is equal to one. To show that, let us consider the first element $(\mathbf{Nx})_1$ of the vector $\mathbf{Nx}$ where every element of $\mathbf{x}$ is constant (i.e. $\mathbf{x}_1 = \mathbf{x}_2 = ... = \mathbf{x}_n = x$)

☕ We obtain by summing all the elements

$$(\mathbf{Nx})_1 = n_{11}x_1 + ... + n_{1n}x_n$$
$$= \frac{1}{k_i}\left(a_{11} + ... + a_{1n}\right)x$$
$$= \frac{k_i}{k_i}x = x = \mathbf{x}_1 \tag{2.71}$$

and similarly for the other components. Therefore $\mathbf{x}$ is really an eigenvector with eigenvalue equal to one. More generally the eigenvalues lie in the range between -1 and 1. In particular, the eigenvalue equal to 1 is always present, with a degeneracy equal to the number of connected components in the graph (similarly to the 0 eigenvalue of the Laplacian matrix), while the eigenvalue -1 is present if and only if the graph is bipartite. Although the normal matrix *is not symmetric*, it has real eigenvalues because it possesses

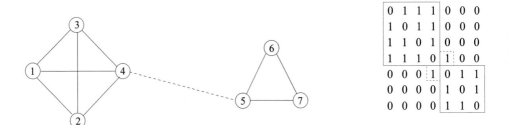

Fig. 2.11 Left: a very simple graph, with two clear communities formed by graphs of order 3 and 4. The edge connecting them is considered as a small perturbation in the space of the subset eigenvectors. Right: the aspect of the related adjacency matrix.

the same eigenvalues as the symmetric matrix $\mathbf{K}^{-1/2}\mathbf{A}\mathbf{K}^{-1/2}$. For the same reasons of normalization valid for the matrix $\mathbf{N}$, in the case of the Laplacian matrix $\mathbf{L}$ we have also the special eigenvalue 0. Formally, the existence of the null eigenvalue can be immediately proved by noticing that all rows (and thus columns) in $\mathbf{L}$ sum up to zero.

When the communities are made by separate and distinct subgraphs the adjacency matrix (as well as the other matrices) is made of distinct blocks. Every block represents the adjacency matrix for a particular subgraph. In this situation, all the subgraphs have a constant eigenvector. The eigenvector resulting for the whole graph is given by composing the different eigenvectors for the subgraphs.

The situation remains similar if we introduce a few edges to connect the subgraphs, as indicated in Fig. 2.11. More generally, in a network with m well-defined communities, matrix $\mathbf{L}$ has also $(m-1)$ eigenvalues close to zero. The eigenvectors associated with these first $(m-1)$ non-trivial eigenvalues also have a characteristic structure. The components corresponding to nodes within the same cluster have very similar values x_i. This means that, as long as the partition is sufficiently sharp, the profile of each eigenvector, sorted by components, is step-like. The number of steps in the profile corresponds again to the number m of communities.

Let us see how this method works by means of a simple example. Consider for example a graph made up of two disjoint complete subgraphs of order 3 and 4 as depicted in Fig. 2.11. The rows of the matrix can be arranged in order to show the presence of these two independent blocks: one 3×3 block and one 4×4 block. The null eigenvalue is then present with twofold degeneracy and the corresponding vectorial space may be spanned by two orthogonal eigenvectors of the type:

$$
\begin{pmatrix} 1 \\ 1 \\ 1 \\ 0 \\ 0 \\ 0 \\ 0 \end{pmatrix} \quad \text{and} \quad \begin{pmatrix} 0 \\ 0 \\ 0 \\ 1 \\ 1 \\ 1 \\ 1 \end{pmatrix} \tag{2.72}
$$

Next we add one edge (the dashed one shown in Fig. 2.11) between two vertices lying in two disjoint components. The result is a matrix slightly changed by a small perturbation. From matrix perturbation theory it can be shown that this perturbation removes the degeneracy of the null eigenvalue (in fact we have only one connected component in the graph now), while at the zero-th order the previous two eigenvectors change into a linear combination of them. There are now two eigenvalues with similar values: the first one is null and the other one is close to zero (because the perturbation was small). Eigenvectors are given respectively by:

$$
\begin{pmatrix} 1 \\ 1 \\ 1 \\ 1 \\ 1 \\ 1 \\ 1 \end{pmatrix} \quad \text{and} \quad \begin{pmatrix} a \\ a \\ a \\ -b \\ -b \\ -b \\ -b \end{pmatrix} \tag{2.73}
$$

with $a = 4$ and $b = 3$ to preserve orthogonality with the trivial eigenvector. By observing the structure of the first non-trivial eigenvector we are able to discern the community structure of the graph. *Vertices are in the same community if their corresponding components in the eigenvector have the same or almost the same value.* The bisection method (Pothen, Simon, and Liou, 1990) works by studying the second eigenvector of the Laplacian (usually computed by means of Lanczos methods (Golub and Van Loan, 1989). We can partition the vertices according to whether their value in this vector is smaller or greater than a splitting value. This method recursively splits the graph into two and therefore in most situations the results are not satisfactory. On the other hand, the situation remains substantially unaltered if more than two clear communities are present. If we have m well-defined communities in the graph, the zero-th approximation is a linear combination of m vectors similar to those in eqn 2.72. The m final eigenvectors have the structure of eqn 2.73. Vertices are in the same community if their corresponding components in the m eigenvectors have a similar value. In this case one possibility is to combine information from the first few eigenvectors, and to extract the community structure from correlations between the same components in different eigenvectors (Capocci *et al.*, 2005). This method is then very similar to the agglomerative methods already seen in the case of edge analysis. In this case the various correlation measures are computed on eigenvector components rather than directly on vertices.

2.6.5   Minimization and communities

The eigenvalue problem can be reformulated in the form of a suitably constrained minimization problem. We consider the most general case and focus on the weighted adjacency matrix $\mathbf{A^w}$, whose elements a_{ij}^w give the strength of the edge between i and j. Let us consider the following function

$$z(\mathbf{x}) = \frac{1}{2} \sum_{i,j=1}^{S} (x_i - x_j)^2 a_{ij}^w \,, \tag{2.74}$$

where x_i are values assigned to the nodes, with some constraint on the vector $\mathbf{x}$, expressed by

$$\sum_{i,j=1}^{S} x_i x_j m_{ij} = 1 \,, \tag{2.75}$$

where m_{ij} are elements of a given symmetric matrix $\mathbf{M}$.

The stationary points of z over all $\mathbf{x}$ subject to the constraint of eqn 2.75 are the solutions of

$$(\mathbf{K^w} - \mathbf{A^w})\mathbf{x} = \mu \mathbf{M}\mathbf{x} \,, \tag{2.76}$$

where $\mathbf{K^w}$ is the weighted degree matrix, $\mathbf{A^w}$ is the weighted adjacency matrix, and μ is a Lagrange multiplier.

Different choices of the constraint M lead to different eigenvalues problems. For example:

- choosing $\mathbf{M} = \mathbf{K^w}$ leads to the eigenvalues problem

$$(\mathbf{K^w})^{-1}\mathbf{A^w}\mathbf{x} = (1 - 2\mu)\mathbf{x} \tag{2.77}$$

- while $\mathbf{M} = \mathbf{1}$ leads to

$$(\mathbf{K^w} - \mathbf{A^w})\mathbf{x} = \mu \mathbf{x}. \tag{2.78}$$

Thus $\mathbf{M} = \mathbf{K^w}$ and $\mathbf{M} = \mathbf{1}$ correspond to the eigenvalue problem for the (generalized) normal and Laplacian matrix respectively.

Thus, solving the eigenvalue problem is equivalent to minimizing the function in eqn 2.74 with the constraint written in eqn 2.75. The x_is are eigenvector components. The absolute minimum corresponds to the trivial eigenvector, which is constant. The other stationary points correspond to eigenvectors where components associated with well-connected nodes assume similar values.

3. Scale-invariance

Fig. 3.1 One of the most renowned fractals; a portion of the Mandelbrot set. The set is defined as the portion of the complex space where (for certain values of the parameter c) the iteration $z_i = z_{i-1}^2 + c$ remains finite.

Many years later, as I faced the editor of this book, I was to remember that distant afternoon when my tutor took me to discover fractals. Even if this episode is rather distant in time, I still remember the feeling of surprise and emotion in front of these objects. This personal experience could perhaps explain the success of fractals in modern science. It is almost impossible to present here a detailed analysis of the various scale-invariant systems appearing in nature. Therefore we give only a brief overview of the successes achieved by using the concept of scale-invariance. Many of the same ideas can be applied directly on self-similar networks (and that is the reason for such a chapter) while other cases are more specific.

The study of scale-invariance has a long tradition. Among the first fields where this property was been analysed were the theory of critical phenomena (Stanley, 1971), percolation (Stauffer and Aharony, 1992), and fractal geometry (Mandelbrot, 1975). Fractal geometry is related to the activity of the mathematician Benoit Mandelbrot[14] (1975, 1983); the well-known Mandelbrot set is shown in Fig. 3.1. One of the first examples considered was the price fluctuations of cotton in the commodities market (Mandelbrot, 1963). The future price cannot be obtained with arbitrary precision from the past series. Yet these series have some form of regularity. The curves for daily, weekly, and monthly price fluctuations are statistically similar. This example is particularly interesting since it clarifies the degree of prediction that can be expected from fractal analysis. Actual prediction is impossible, but we can measure the probability of some future events. The fact that the same statistical features are found on different timescales is a typical sign of fractal behaviour. Similarly, in the case of coastline lengths we find fractal objects. If we try to measure the total length, the real shape is so complicated that we always miss parts of the profile even by using an infinitely small ruler. Actually fractal behaviour might refer to different properties. In some systems the scale-free structure is in the shape. In this class the fractal shape can be 'robust' as is the case of the branched patterns of dendrite growth, fractures, and electric breakdowns. We say robust because these phenomena happen for a variety of external conditions. In the same class we have other systems that are more 'fragile'; fragile in the sense that they arise after a very precise tuning of some physical quantity. This is the case for percolation and critical phenomena. For a completely different class of phenomena the scale-invariance is not geometrical but related to the dynamics or the evolution of the system. For example, the time activity of a system could display self-similar behaviour, in which case inspection by eye would not help. The only possible sign of fractal behaviour is the mathematical form (power-law fluctuations) of the time series. This happens for connected events of temporal activity, also known as avalanches. A series of aftershocks after a principal earthquake forms an avalanche; a series of consecutive topplings in a granular system forms an avalanche; and in general any series of response of the system to an external perturbation forms an avalanche. Very often the probability distributions of energy released in an avalanche of fracture (for example) as well as the duration (or size) of the

---

[14] Born in Warsaw in 1924, he moved to France in 1936. He is currently Sterling Professor of Mathematical Sciences, Mathematics Department, Yale University and IBM Fellow Emeritus at T.J. Watson Research Center.

avalanche follow a power-law distribution. We regard these phenomena as
fractals in time.

Finally self-similarity can be present in the way the different parts of a
system interact with each other. This is the case for self-similar graphs and
the power-law scaling appears in the distribution of topological quantities
like the number of interactions per part of the system. We regard these
phenomena as fractals in the topology. To detect when and if this happens
is the main topic of this book.

> In this chapter we present some experimental evidence of
> scale-free structures. The most immediate of them are geometri-
> cal objects that result from mechanical or electrical breakdown.
> They are 'traditional' fractals. Other scale-free relations present
> in nature involve completely different quantities as probability dis-
> tributions or time activity. All of them are characterized by the
> same shape and sometimes we can find similar causes for this fea-
> ture. Finally, we show how to measure fractal dimension and more
> generally how to plot power laws.

3.1 Geometrical scale-invariance: fractals

Self-similarity describes invariance of physical objects with respect to a
change of the scale of observation. From very basic geometry, we know that
the technical term 'similarity' is introduced to describe objects (for example, two triangles) whose corresponding edges are proportional. Whenever
this happens, we know that by enlarging the smaller object by this factor of
proportionality we obtain a perfect copy of the larger one.

Fractals are more complicated than triangles and have the property of
being similar to themselves. This means that a small part of them, when
enlarged, is a copy of the whole. It is then natural to refer to such systems as
self-similar. Even if this not a rigorous definition, we can define empirically
the self-similar (hereafter *'fractal'*) objects, as those that look the same
when decreasing or increasing the scale of observation. Probably the most
famous object of this kind is the Mandelbrot set shown in Fig. 3.1 and on the
right of Fig. 3.2. The shape of the set is determined by exploring which set of
complex numbers has the property of keeping a certain function finite.[15] The

---

[15] The Mandelbrot set is a region of the complex space corresponding to the numbers
c such that the succession $z_n = z_{n-1}^2 + c$, (where $z_0 = 0$) remains finite.

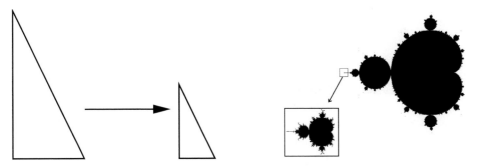

Fig. 3.2 The concept of self-similarity. On the left, two triangles; they are similar each other. On the right, the Mandelbrot set. A subset of this system is similar to the whole.

region of the complex space formed by such numbers forms these incredibly complicated lines of separation. Using this idea, we now try to build a fractal object trying to provide step by step a more rigorous characterization of it.

3.1.1   Fractals by iteration

The procedure is shown in Fig. 3.3. We start with a triangle whose size is L; we then divide it into four parts and remove the central one. This produces the second object in the figure. Now we iterate this recipe for each of the three surviving triangles, obtaining the third object in Fig. 3.3. It is easy to check that any of the subsets with a size of half of the system resembles the original one at scale one.[16] By abstraction we can imagine repeating this procedure infinitely, obtaining an object that would look very similar to the last object in Fig. 3.3 (actually composed by only five iterations).

This object is known as Sierpinski Gasket[17] and we now characterize it in a more precise way. The striking difference with respect to the starting triangle is the regular presence of empty regions between the black ones. This strange distribution of filled and empty regions makes it impossible to measure the area (how much space is filled) of the objects. As a first approximation, we see that the area of the compact triangle is an overestimate of the area of the Sierpinski Gasket. To find the right value we must generalize the usual concepts.

---

[16] Technically that is true provided one cuts off the scales of observation, i.e. considering only sizes greater than 1/8 and smaller than 1.

[17] Named after the mathematician Waclaw Sierpinski, born in Warsaw in 1882 and died in the same place in 1969.

Fig. 3.3 Starting on the left we have three steps of construction of the Sierpinski gasket. After an infinite number of steps it appears as the last picture on the right (actually only five steps of iteration).

Measuring something consists in comparing it with a sample object called a 'unit of measure'. When we say that a building is ten metres high we mean that we would need to stack ten times one metre blocks on top of one another in order to have the same height. The same procedure holds also for surfaces: an area of 10 square kilometres is covered by 10 replicas of a large square of 1 kilometres per side (even if this is not the smartest way actually to make such a measurement). A less trivial question arises if we want to know how many unit measures of surface we need in order to cover a Sierpinski gasket. For the first iteration we only need three triangles of side $1/2$ (Note that for a whole triangle we would need four of them.) For the second iteration we need nine triangles of side $1/4$ (Note again that for the whole triangle we would need sixteen of them). In general for the compact triangle the number of triangles needed grows quadratically as we reduce the size. This idea is at the basis of the definition of the **fractal dimension** D.

✥ More formally, if $N(\epsilon)$ is the number of triangles of size $1/\epsilon$, then the fractal dimension is defined as

$$D = \lim_{\epsilon \to 0} \frac{\ln N(\epsilon)}{\ln 1/\epsilon}, \tag{3.79}$$

which is constant as ϵ changes and gives a measure of the empty regions present in a fractal.

For a compact object the definition of fractal dimension gives the Euclidean dimension (for a compact triangle we have $D = 2$). This is because for the first iteration the number of triangles is 4 and $1/\epsilon$ is 2, therefore $D = \frac{\ln 4}{\ln 2}$. When passing to the next iteration $1/\epsilon$ is 4 and the number of triangles is 16 so that again $D = \frac{\ln 16}{\ln 4} = 2$. It is easy to understand that we always find the same value of D for all subsequent iterations (i.e. in the limit $\epsilon \to 0$). So far, this is only a very complicated way of saying that a triangle has 'Euclidean' dimension $D = 2$.

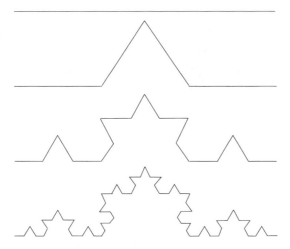

Fig. 3.4 The Koch curve, obtained by infinite recursion of kink production in the middle part of the segment.

Things become more interesting by considering the fractal Sierpinski Gasket. By using the previous formula, we obtain at the first iteration that

$$D = \lim_{\epsilon \to 0} \frac{\ln N(\epsilon)}{\ln 1/\epsilon} = \frac{\ln 3}{\ln 2}. \qquad (3.80)$$

Again the value of D does not depends on the particular ϵ, since at the next iteration we have $D = \frac{\ln N(\epsilon)}{\ln 1/\epsilon} = \frac{\ln 9}{\ln 4} = \frac{\ln 3}{\ln 2}$. The value of D is no longer an integer since $D = \ln(3)/\ln(2) \simeq 1.58496$. This quantity called **fractal dimension** measures the difference of compactness of a Sierpinski gasket with respect to a regular triangle. The fractal dimension D is lower than 2 because the Sierpinski Gasket is less dense than compact figures. D is also larger than 1 and that is reasonable because the gasket is denser than a line.

This is only one example. Different objects can be produced in a similar fashion. Here we produced a fractal by iteration, forming more and more empty regions in a compact truly two-dimensional triangle. We can produce fractals with the opposite procedure. That is, we can kink a single segment, a truly one-dimensional object, and obtain a qualitatively new geometrical figure.[18] Starting with a line segment, we divide it into three different pieces and replace the inner one with a reversed V shape as shown in Fig. 3.4. As usual, we must imagine how the system appears after an infinite series of

---

[18] This clarifies the etymology of the word 'fractal'. B. Mandelbrot coined this word from the Latin adjective *fractus*, i.e. broken.

iterations. The final shape is somewhat similar to the last picture in Fig. 3.4. By computing now the value of the fractal dimension we find a value of

$$D = \lim_{\epsilon \to 0} \frac{\ln N(\epsilon)}{\ln 1/\epsilon} = \frac{\ln 4}{\ln 3} \simeq 1.2618, \tag{3.81}$$

which is denser than the value expected for a simple linear object.

3.1.2   Stochastic fractals, scaling and self-affinity

Passing from mathematics to applied sciences the situation changes a little bit. Real data do not look so regular and nice as the pictures presented above. No iterative formula is given for the case studies. Therefore, all the relevant information on fractal dimension must be extracted from the data. In textbooks the difference between toy objects like the Sierpinski Gasket and real fractals is sometimes indicated as the difference between 'Deterministic Fractals' and 'Stochastic Fractals' respectively. That is not completely correct since you can also have statistical randomness in the toy pictures like the Sierpinski gasket. For example, a stochastic toy fractal is a Sierpinski Gasket where we remove the inner triangle with probability p.

Those stochastic fractals are more similar to the real ones. We cannot expect that for any small division exactly the same recursion takes place. The most probable case is that *on average* the number of boxes needed to cover the structure peaks around a mean value with occasional oscillations to smaller or larger values. Since we are interested in the limit behaviour when the linear size of the box tends to zero we have to infer the limit behaviour of this averaged quantity (see next section).

Another important extension of self-similarity is given by the concept of *self-affinity*. An affine transformation in geometry consists in applying different magnifications along different directions. For instance, we can pass from a square to a rectangle through an affine transformation. As in the case of self-similar objects, self-affine objects are 'affine' to their subsets. In the following we present different examples of quantities whose magnification properties are different along different directions. We refer to this property by saying that they have different 'scaling' properties. This means that if the object is doubled along one direction it could be three times longer along another. The two different directions have a different 'scaling'.

A typical curve that shows self-affine characteristics is the price curve $p(t)$ of a stock. This curve is determined by plotting on the y-axis the price of a certain stock at time t (reported on the x-axis). Since the function is single valued (which means that for any value of time t there must be one and only one price) oscillations are allowed only on the y-axis. Another way to measure such a property is to measure the *roughness* of the curve by

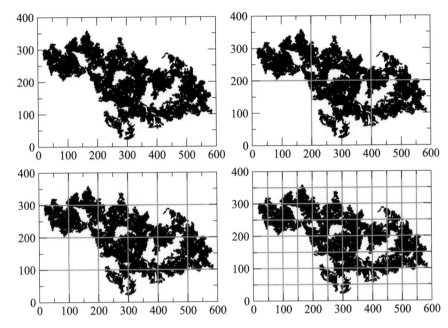

Fig. 3.5 The first steps of box-counting for a satellite image of a wildfire. Every black pixel is a portion of burned area.

means of the roughness exponent α appearing in the scaling formula:

$$p(t) \simeq b^{-\alpha}p(bt) \qquad (3.82)$$

that gives the order of magnitude of price fluctuations in time. Self-affinity is a generalization of the concept of self-similarity (which is no longer appropriate), and also in this case the mathematical form of the relationship is that of a power law.

3.2   Measuring the fractal dimension

There are two main procedures for measuring the value of the fractal dimension from experimental data. While in principle they are both based on the concepts explained in Section 3.1.1, in practice one or the other could be more appropriate to the case of study.

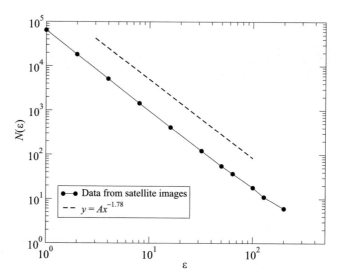

Fig. 3.6 The plot of $N(\epsilon)$ versus ϵ for the case of wildfires shown in Fig. 3.5.

3.2.1   Box-counting

This method called '*Box-counting*', follows immediately from the mathematical definition of fractal dimension. As shown in Fig. 3.3 and mathematically expressed in eqn 3.79 and eqn 3.80, the idea is to count the portion of the surface covered by the fractal (this generalizes immediately to one and three dimensions). This method can be applied whenever we deal with a picture or a spatial grid of data. Let us consider the case of Fig. 3.5, where a satellite image of a wildfire is shown. The method counts the number of units of measure (in this case boxes) needed to cover the structure. Of course two limit situations are present. Boxes much larger than the whole set are uninteresting, the structure is inside in any case. This is called the 'upper cut-off'. This is not useful information of course, and this is the reason why the fractal dimension is often defined when the linear size of the boxes goes to 0. Unfortunately, for a physical object there is also a lower cut-off given by the image resolution. This means that it is useless to consider units smaller than a pixel of the image. Then one simply plots the number of boxes occupied by the structure when the size of this box passes from one (as large as the whole set) to the size of the pixel of the image. Different stages of this procedure are presented in Fig. 3.5, while the fit for the fractal dimension is shown in Fig. 3.6.

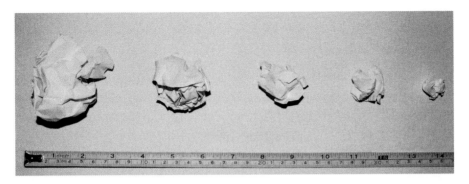

Fig. 3.7 One realization of the experiment. A ruler with centimetres and inches is shown below.

3.2.2   A home-made fractal, the mass-length relationship

In some cases (e.g. three-dimensional objects), a picture of the fractal set would be not appropriate. To overcome this problem, one can use a different method by considering the mass of the object. In the following we assume that the mass density is uniform across the fractal. Therefore fluctuations in the mass are only given by the geometry of the sample. Since for a compact object we have $M \propto \rho L^D$ (where M is the mass, L its linear size, and ρ is a constant called the 'form factor' related to the shape of the object), we expect the same relation to hold also for a fractal (where D is a non-integer number). Therefore by measuring an object and weighing its mass it is possible to compute its fractal dimension. This method is called the *'mass length relationship'* and we can illustrate it with a little experiment.

Let us take two sheets of paper with A4 format. A good quality printing paper weighs 80 g/m^2, so that an A4 (210 mm × 297 mm) weighs 4.9896 g. Take one of the sheets and crumple it into a ball. Apart the change in its shape, this sheet still weighs about 5 grams. Then take the other sheet and tear in half. Now crumple one of these smaller sheets. It will form a ball weighing about 2.5 grams. Tear the remaining sheet in half and again crumple one of the two pieces. This ball weighs about 1.2 grams. Continue like this for a few more times and you are left with 5 or 6 balls of different masses (see Fig. 3.7). If one now measures the linear size of these objects one obtains approximately the values shown in the inset of Fig. 3.8. The result is that crumpled papers form a fractal whose dimension is $D \simeq 2.5$.

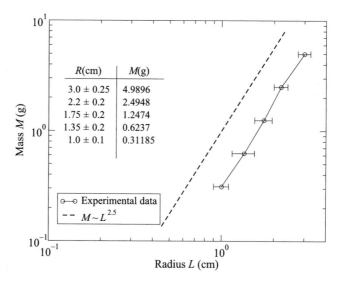

R(cm)	M(g)
3.0 ± 0.25	4.9896
2.2 ± 0.2	2.4948
1.75 ± 0.2	1.2474
1.35 ± 0.2	0.6237
1.0 ± 0.1	0.31185

Fig. 3.8 The mass length relationship for the crumpled sheets allows one to determine the fractal dimension of these objects.

3.3   Scale-invariance and power laws

The mathematical form of self-similarity is represented by power laws. Whenever a function $y = f(x)$ can be represented as a constant to the power of x then the relation between y and x is a power law. The linear function $y = x$ and the quadratic function $y = x^2$ are the simplest examples of power laws. A physical example is the elastic force which grows with the distance r from the equilibrium position as a power law ($F_{el} \propto r$). The gravitational and electrostatic forces F_G and F_E decay with distance as a power law with exponent -2 ($F_G \propto 1/r^2$, $F_E \propto 1/r^2$). In the case of fractals we have seen above that their geometry can be identified by considering the number of boxes $N(\epsilon)$ of linear size $1/\epsilon$ needed to cover the structure. In particular,

$$N(\epsilon) \propto (1/\epsilon)^D \rightarrow N(\epsilon) \propto \epsilon^{-D}. \qquad (3.83)$$

That is, the number of boxes is a power law of linear size with an exponent D, called the *fractal dimension*. The above formula is a way to define D, but this quantity is usually measured by using another power law relation. This is the above-mentioned mass length relationship

$$M \propto L^D. \qquad (3.84)$$

Scale invariance is not restricted to geometry but also appears in dynamical systems. In this case we have power-law distributions for different physical

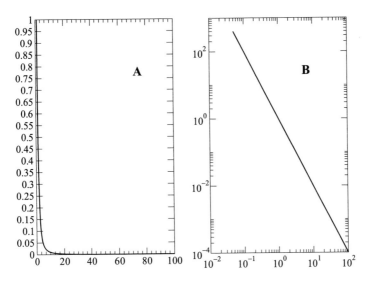

Fig. 3.9 The plot of the function $y = x^{-2}$ in linear (**A**) and logarithmic (**B**) scale.

quantities. For example the evolution of some systems (sandpiles, number of species in an ecosystem) proceeds with a series of causally connected events (avalanches) whose size s is again distributed as a power law

$$P(s) \propto s^{-\tau}. \tag{3.85}$$

3.4   Plotting a power law

Now that we have defined a suitable way of measuring the fractal dimension, the next step is to fit the experimental data. If we plot the data on a double logarithmic scale,[19] we should obtain a straight line, as shown in Fig. 3.9. This is because

$$y = x^{\alpha} \rightarrow \ln y = \alpha \ln x. \tag{3.86}$$

The tail of the distribution shown in Fig. 3.10 **A** is very noisy. This is a general feature in many real experiments. If we realize N experiments, we are likely to sample the interval between 10 and 100 more accurately than

---

[19] Generally a logarithm is defined by assigning a basis, therefore $b = log_a c$ means that $a^b = c$. Unless specified differently the basis of logarithms we consider is the number $e = 2.7182818284 \dots$ This is indicated by symbol ln. Figures are always represented with base 10 logarithms (indicated by symbol log). We can pass from one basis to another by a simple multiplicative factor.

the interval between $1,000$ and $10,000$. Nevertheless both intervals are of the same size on a logarithmic scale. Since N is finite for a real experiment, we cannot avoid noise in the tail of the plot. Different solutions are possible to overcome this problem. We show them in Fig. 3.10 **B** and Fig. 3.10 **C**. In the first case we applied the method of 'binning' while in the second case we used a cumulative distribution.

These methods make use of the fact that the sum of fluctuations from average must be zero for statistical noise.[20] When summing the data, the fluctuations cancel and we recover the average. An important difference between Fig. 3.9 and Fig. 3.10 is that in an experiment we cannot have frequencies lower than 1 (we cannot measure one event 0.3 times). This lower cut-off will play an important role in the following sections.

3.4.1   Binning procedure

In this method the noise reduction is done by dividing the x-axis into intervals called bins and averaging the data within each bin. As an example, we can take the frequency of all the numbers between 1 and 10. Let us assume that the bin is 10 units wide: we can represent all the data in one point b with x_b given by the average of the bin extremes, and y_b given by the average value of ys. If the bin size is constant, for large values of x the density of bins becomes very large as well. Furthermore, as noticed above, as the x grows, more and more trials would be necessary to test any single bin of length 10.

In the case of power laws it is useful to adopt a *logarithmic binning*. For example, take the size of the first bin to be two (i.e. average all the points between 1 and 3), then the second bin will have a size of four (i.e. between 3 and 7), the third one of eight (i.e. between 7 and 15) and so on. In this case the size of the bin is a power of 2 (i.e. $2^1, 2^2, 2^3...$) A possible choice is to take as y_b the average of the values ys in the bin and as x_b the geometrical average of the bin extremes. In this way the bins are equally spaced on a logarithmic scale. Note that the basis can be any number larger than 1. For instance in Fig. 3.10 **B** we used bins whose size is given by powers of 1.2. The drawbacks of this procedure (Newman, 2005) are the following: this method does not completely reduce the noise and the choice of most appropriate bin size must be determined by using trial-and-error. In general, for a small and noisy set of data, interesting behaviour in the tail of the distribution will be lost if the bins are too wide. On the other hand, bins that are too narrow will not succeed in averaging the fluctuations.

---

[20] This is very simple to see, in formula the sum of fluctuations from average is given by $\sum_{i=1,N}(x_i - \langle x \rangle) = N\langle x \rangle - N\langle x \rangle = 0$.

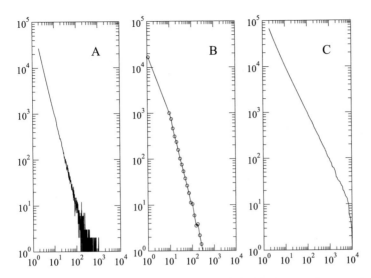

Fig. 3.10 A The plot of a computer-generated (with imposed power-law shape) frequency distribution of random numbers. **B** The plot of the same frequency distribution done with bins of exponential size. The first bin is between 1 and 1.2, the second between 1.2 and 1.2^2. In general the n-th bin collects data between 1.2^{n-1} and 1.2^n. **C** The cumulative distribution of the same data. Note the change in the slope.

3.4.2   Cumulative distribution

This method works by averaging out fluctuations across the entire data set. Instead of determining the probability that a certain value x (or in the case of a continuous formulation a value between x and $x + dx$) appears in the experiment, we focus on the probability $P^>(x)$ that an outcome is *larger* than x.

 This corresponds to introducing a new probability distribution $P^>(x)$ given by

$$P^>(x) = \int_x^\infty P(x')dx'. \qquad (3.87)$$

In this case, if the $P(x)$ is a power law of the kind $P(x) = Ax^{-\gamma}$, this again gives a power law

$$P^>(x) = \int_x^\infty P(x')dx' = \int_x^\infty Ax'^{-\gamma}dx' = \frac{A}{\gamma - 1}x^{-\gamma+1}. \qquad (3.88)$$

The 'cumulative distribution' $P^>(x)$ is still a power law but with a modified exponent $-\gamma + 1$. The results are shown in Fig. 3.10 **C**, where most of the noise has been greatly reduced and the slope of the distribution has changed according to eqn 3.88. Note that despite the widespread use of this method one can sometimes experience catastrophic results:

- In principle, if the exponent γ is close to one, the integral does not behave like a power law but rather like a logarithm.
- Almost always the upper limit of integration is a finite value x_{max}. This has the effect of considerably changing the shape of the curve.

 The correct integration of equation 3.88 would then give

$$P^>(x) = \int_x^{x_{max}} Ax^{-\gamma}dx = \frac{A}{\gamma - 1}(x^{-\gamma-1} - x_{max}^{-\gamma-1}). \qquad (3.89)$$

The effect is that the cumulative distribution will resemble a power law more and more closely only as the value of x_{max} tends to infinity. If x_{max} is too small, the deviation of the distribution from a straight line could make estimating the value of γ very difficult.

3.5   Scale-invariance in natural sciences

The most widely known family of fractal structures in natural sciences is given by the class of fractals related to deposition phenomena. In electro-deposition (Brady and Ball, 1984), a crystalline deposit of metal grows on one of the electrodes put in a solution of metal ions. This process takes place when a difference of electric potential is applied to the electrodes. When this difference is reasonably low, ions have the time to find their way to the electrode and the deposition results in a highly packed sample. Moreover when a higher potential is applied, the growth is quicker and such an optimal packing is not obtained. In this case, they show a characteristic branching pattern.

It is interesting to note that the same branched patterns are also observed in other phenomena; for example, in the patterns generated by the diffusion of manganese ions through cracks inside rocks. Whenever manganese ions come in contact with oxide ions of the rock they form a black compound (MnO) that precipitates if the concentration is larger than a certain threshold. Presence of the crystallized compound triggers new precipitation (as in the DLA irreversible attachment), forming the typical patterns (Chopard, Herrmann, and Vicsek, 1991). When the deposition surface is smooth, further growth is equally likely to occur anywhere on the surface. However once small bumps start to form, successive deposition is more likely to take place there, resulting in the formation of branches. These branches restrict the

Fig. 3.11 The familiar pattern of a lightning (Courtesy of Bert Hickman, Stoneridge Engineering, http://www.teslamania.com).

freedom of movement for the molecules in the solution. As a macroscopic effect, the growth only happens at the tips of branches. This basic mechanism takes place in a variety of different physical scenarios, for instance in electric discharge. A dielectric is a material that does not conduct electricity. This means that the electrons of such a material are not free to move in the sample. When the electric field applied to it exceeds a certain threshold this property is no longer valid. Rather, the electric field is now strong enough to attract the electrons out of their atoms, breaking the bonds of the material. The dielectric breaks up and a discharge flows through it. In air this is the phenomenon of lightning. Assuming that breakdown takes place with a probability proportional to the strength of the applied electric field (Niemeyer, Pietronero, and Wiesmann, 1984), one can reproduce the statistical features of real lightning as shown in Fig. 3.11. As for deposition, here too growth is irreversible and starts from an initial failure, forming branches, since the electric field is stronger at the tips. The same happens when the driving field is represented by concentration pressure. The tip instability is called Saffman-Taylor instability and it underlies the viscous fingering phenomenon, as shown in the left part of Fig. 3.12. Even bacterial growth can be described in this way, where the gradient is given by the concentration of food (Matsushita and Fujikawa, 1990) (see right part of Fig. 3.12).

We can approach the problem of fractures in the same spirit. Whenever a solid object is subjected to an external load, it experiences a change of shape. The situation is complicated by the vectorial nature of the problem. As is evident in the case of a rubber string, an elongation (i.e. a deformation

Fig. 3.12 Left: a pattern of viscous fingering, courtesy of M. Thrasher (Ristroph *et al.*, 2006). Right: colonial developments of *Paenibacillus dendritiformis*, courtesy of K. Ben Knaan and E. Ben-Jacob (Ben-Jacob and Levine, 2001).

along its length) is very often[21] coupled with a diameter reduction (i.e. a deformation orthogonal to the force applied). When the force applied is too large, the solid eventually breaks apart. Very often the surface of the fracture shows a fractal shape. Much research has been done on the characterization and modelling of the various cases of fracture, but we shall not enter into the details of fracture theory here. Nevertheless, by using highly simplified models of breakdown, we can spot a behaviour similar to electric breakdown. This is because also in fractures the load is highest at the tips of the crack.

3.5.1   Scale-invariant systems in physics: percolation and critical phenomena

3.5.1.1   Percolation   The original model of **percolation** is one of the most ancient and elegant models of fractal growth. It has been introduced to explain the phenomenon of fluid penetration in a porous medium. The idea is that a cluster of percolation is composed by points of a lattice extracted with a certain probability p. When $p = 0$ there are no points in the cluster; when $p = 1$ all the points of the lattice belong to the cluster. Interesting phenomena happen for intermediate values of p. When this occupancy

---

[21] With the notable exception of corks that can be stretched with no side deformation and two-dimensional hornbeam leaves (Mahadevan and Rica, 2005) that widen when they are elongated.

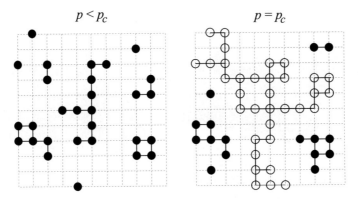

Fig. 3.13 Left: an example of a system where nodes are drawn with a probability lower than the critical one p_c. Right: at the critical value of percolation, new sites are drawn and a percolating cluster (white sites) with fractal properties appears in the system.

probability grows from 0 to 1 small clusters of points appear in the system. At a precise value $p = p_c$, depending on the particular lattice considered, a giant fractal cluster spans the system from one side to the opposite one (see Fig. 3.13). A self-organized version of this model of growth has more recently been introduced (Wilkinson and Willemsen, 1983) and it is called *invasion percolation*. It was introduced in order to explain the patterns observed when pushing water in a medium containing oil. The medium is represented by a regular series of bonds between the sites of the lattice. Every bond is assigned a random variable describing its diameter. Because of the capillarity effect, the invasion of water proceeds by selecting the bond with the smallest diameter on the boundary of the region invaded by water. This model self-organizes in a critical state, forming a fractal percolation cluster in the steady state. This feature is particularly interesting since self-organization could represent a possible explanation for the ubiquity of fractal structures (see Section 4.2.1). There are many real situations where percolation models can be used to describe experimental results. Here we report the study of wildfires from satellite images. From these pictures it is easy to note that after a wildfire the area of vegetation burnt has a typical fractal shape (see Fig. 3.14). Various explanations have been provided for such phenomenon, most of them closely related to the idea of percolation.

The particular data sets shown in Fig. 3.14 consist of Landsat TM satellite imagery (30 m ×30 m ground resolution) of wildfires, acquired over the Biferno valley (Italy) in August 1988 and over Mount Penteli (Greece) in July 1995, respectively. In all the cases the image was acquired a few days

Fig. 3.14 Left: a wildfire originated in Valle del Biferno (Italy) in August 1988 destroying 56 square km of vegetation. Right: a wildfire on Mount Penteli (Greece) that in July 1995 destroyed 60 square km of vegetation.

after fire. The burnt surfaces were respectively 58 and 60 square kilometres. Bands TM3 (red), TM4 (near infrared), and TM5 (mid infrared) of the post-fire sub-scene are classified using an unsupervised algorithm and eight *classes*. This means that in the above three bands any pixel of the image is characterized by a value related to the luminosity of that area. By clustering those values in *classes* one can describe different type of soil, and in particular the absence or presence of vegetation. In particular, the maps of post-fire areas have been transformed into binary maps where black corresponds to burned areas. Time evolution of wildfires is better described by a dynamical version of percolation in which the probability of ignition decays with time so that at the beginning the fire will grow almost in a compact way leaving a fractal boundary at the end of the activity.

3.5.1.2   Critical phenomena   Percolation is not the only process that forms a fractal by a careful tuning of a physical quantity. Traditionally this is the case for critical phenomena in thermodynamics. Consider a typical pure (for our purposes we restrict to only one type of molecule of a simple chemical element) chemical substance. Then plot on a graph the values of pressure and temperature at which this substance changes its phase. This means when it becomes solid from liquid (solidification) or vapour from liquid (evaporation) or vapour from solid (sublimation).[22] In general, all the information can be reported in a chart like the one presented in Fig. 3.15. The fact that the vapour liquid curve stops means that if we choose the

----

[22] These transitions can also happen at the same temperature and pressure in the other direction (in this latter case they take respectively the names of melting, condensation, and deposition).

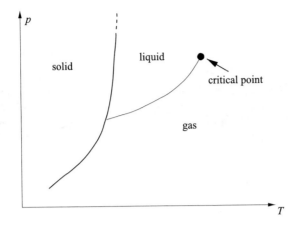

Fig. 3.15 A typical phase diagram for a pure substance, the left (thicker) line gives the separation between liquid and solid at the various pressures and temperatures. Notably water has a different (and much more complicated) phase diagram. For example, in a certain range of the parameters the slope of this line is negative rather than positive, so that at fixed temperature an increase of pressure liquefies the ice (this phenomenon allows skiing and skating).

right path we can go from one phase to another without having a phase transition. This is a peculiar situation. In the standard cases (liquid solid or liquid vapour below the critical point) the passage between one phase and the other happens abruptly (we have a jump in some thermodynamical quantities like the specific heat). As the critical point is approached instead, the system adjusts itself on a microscopic level. At temperatures T near the one of the critical point T_c the quantities like the specific heat, the compressibility, etc. are power laws of the quantity $(T - T_c)$. Regardless of the precise substance or variable involved, many systems approaching the critical point present a similar behaviour. This property is known as universality.

3.5.2 Scale-invariance in time activity: avalanches and extinctions

There are some physical situations (as in the case of networks) where the self-similarity does not hold for the geometrical properties. For instance, this happens when the time evolution is characterized by bursts of activity separated by long periods of quiescence. In such cases the *dynamics* of the system is characterized by series of causally connected events called 'avalanches'. One classical example is that of earthquakes. Very often a major event comes along with a series of pre- and aftershocks. In most of the cases the statistical distribution of these processes again follows a power law. In the completely different field of ecology a similar dynamics is believed to be behind species differentiation between plants and animals. From fossil records it appears that during particular moments of the Earth's history such as the Permian (291 million years ago) or the Cambrian (570 million years ago), many different species appear. In other periods a similar size of extinctions was present. This led to the formulation of the theory of *punctuated equilibrium*. In this theory (Eldredge and Gould, 1973) the evolution takes place by means of intermittent bursts of activity rather by means of progressive changes (see Section 8.1.1 and Fig. 8.3).

Many models have been introduced inspired to these ideas. Amongst them the most famous is probably the Bak-Sneppen model (Bak and Sneppen, 1993). Here a food chain in an ecosystem is represented by a series of species arranged on a one-dimensional lattice. A species i is represented by a real number η_i giving its fitness. Species with a low fitness are unlikely to survive in the ecosystem. In the model, the species with the lowest value is removed together with its prey and its predator. Their place is taken by three new species with randomly extracted fitness values. This elementary process represents an avalanche of extinctions of size 3. If one of the newcomers also becomes extinct then the avalanche lasts for another time step. The avalanche can be defined as the number of species going extinct in a series of causally connected events (also the number of time steps gives a similar measure). The process continues with new species entering the system. The size of these avalanches shows no particular timescale, and large-scale extinction events that sweep the entire system are separated by long periods of little activity.

Another way to find scale-invariance is in the distribution of the strengths of events, rather than in their duration. Returning to the case of earthquakes, we again find a power law.[23] The strength of an earthquake is computed through its amplitude A, and statistics are collected for both specific areas and the whole planet. The result is the law of Gutenberg and Richter (1955)

$$P(A) \propto A^{-\gamma}, \tag{3.90}$$

which indicates that no characteristic scale for earthquakes must be expected. In Fig. 3.16 we report the statistics for various databases. This

---

[23] Note that in the case of earthquakes this is coupled with a fractal spatial distribution of epicentres.

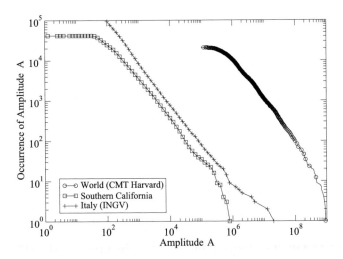

Fig. 3.16 Three different earthquakes data sets (courtesy of P. Tosi). The first was collected by Harvard University using CMT (Centroid Moment Tensor) for world earthquakes. The second is the SCE (Southern California Earthquakes). The third is a collection of Italian earthquakes recorded by the Italian Institute for Geophysics and Vulcanology INGV (Istituto Nazionale Geofisica e Vulcanologia).

fact is well known and earthquakes are often classified according to their magnitude, that is the logarithm of the amplitude.

 Technically, the *magnitude* of an earthquake is defined as the difference between the (base 10) logarithm of the amplitude A of the earthquake (as measured by a device called seismograph) and the logarithm of the amplitude A_0 of a reference earthquake, so that

$$M = \log_{10} A - \log_{10} A_0. \qquad (3.91)$$

The amplitude is measured in millimetres of the elongation in a seismograph; it is supposed to be proportional to the energy released.

3.6   Scale-invariance in economics and in social sciences

One of the first examples of fractal geometry was related to commodity prices (Mandelbrot, 1963). In particular, a chart of the different prices of cotton in a trading period look similar to another chart with a different time resolution.

Actually, power laws in economics were introduced even earlier, by the seminal and pioneering work of Vilfredo Pareto.[24] Pareto noticed that in a variety of different societies, regardless of countries or ages, the distribution of incomes and wealth follows what is now called *Pareto's law* (Pareto, 1897)

$$N(X > x) \propto x^{-m} \tag{3.92}$$

where $N(X > x)$ is the number of income earners whose income is larger than x.

A similar law can be derived in a completely different context. In linguistics for example, one may be interested in the frequency of use of different words in a text. In this case scale-invariance appears again under the form of a power law. This result called *Zipf's law* (Zipf, 1949), after George Kingsley Zipf,[25] states that

$$f \propto r^{-b} \tag{3.93}$$

where f is the frequency and r is the rank of the word (from the most to the least used). Since b is in most cases equal to one, we have that the frequency of a word is roughly inversely proportional to its rank. Zipf's law is far more general than linguistics, the above case applies to city populations as well as species abundance. More generally, Zipf's and Pareto's laws are two ways to represent the same phenomenon. Whenever we deal with a power-law density distribution both Zipf's and Pareto's laws apply. We can relate the value of the exponents m and b (Adamic, 2002). The idea is that when an object whose rank is r has frequency y, it means that r different words appear more than y times. This is the prediction of Pareto's law, apart from the fact that axes x and y are inverted. So that

$$f \propto r^{-b} \rightarrow r \propto f^{-1/b} \tag{3.94}$$

Now the frequency f is analogous to incomes x as rank r is analogous to the number of people with income larger than x and therefore we obtain for the two exponents the following relation

$$1/b = m \tag{3.95}$$

In the original formulation of Pareto's law, we were considering the cumulative distribution of incomes rather than their density. As we recall in Appendix C, the first is the integral of the second. This means that the

[24] Italian economist, born in Paris in 1848 and died in 1923 in Céligny, Switzerland.
[25] US Professor of Linguistics at Harvard (1902-50).

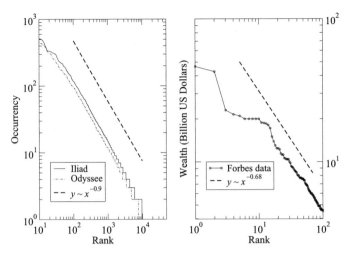

Fig. 3.17 Left: Zipf's law for the Greek versions of Odyssey and Iliad. Right: the income distribution for the 100 wealthiest persons in the world. We use here the Zipf form of the Pareto distribution for a comparison between the two phenomena.

frequency with which we have persons with income exactly x (in the case of continuous functions we need a dx) is

$$N(X = x) \propto x^{-m-1} = x^{-a}. \tag{3.96}$$

So when considering the probability functions instead of the cumulative we have the exponent relation

$$1 + 1/b = a \tag{3.97}$$

The validity of the above laws can be tested easily for two particular case of studies. In the case of Zipf's law we report the analysis of the text of the Odyssey and Iliad where the different forms a word can have in ancient Greek (declination for nouns and adjectives and conjugation for verbs) have been considered as different words. English has a much more limited number of different forms for a single word, but nevertheless the same texts present a similar frequency distribution. By using the data set provided by Forbes company (http://www.forbes.com) we can test the law of Pareto for the 100 wealthiest persons in the world in the period 2000-4. Apart from the first ten or so who seem to deviate from the expected behaviour, the others show a rather nice power-law behaviour, as shown in Fig. 3.17.

4. The origin of power-law functions

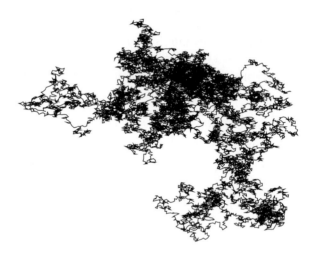

Fig. 4.1 A random walk of one million steps. One portion of the walk has the same complexity of the whole. Diffusion processes are one of the most common examples of fractal growth.

All self-similar systems are power-law alike. Each power law we see in nature is self-similar in its own way. The aim of this chapter is to present various possible reasons why things as different as scale-free networks, geometrical fractals, and avalanche phenomena all behave as power laws. Quite surprisingly amongst the several causes we also find randomness. This means that disorder (fluctuations) in the quantity values between different sub-parts of the system, is enough to produce power laws. Various mechanisms producing power laws are presented in this chapter; further extensive discussion of this topic is available in the literature (e.g. Mitzenmacher 2004). A summary of the topics of this chapter is as follows:

- *Diffusion processes.* The second law of thermodynamics states that things tend to mix. For example, particles of one kind diffuse into particles of another kind. Similarly, heat tends to be transferred from one body to another through the scattering of atoms. The mechanism of diffusion is a powerful power-law generator, as we see in Fig. 4.1.
- *Minimization.* Some recent trends try to show that self-similarity of any kind could be related to a common evolutionary process. This is in some sense related to other theories predicting that self-similarity arises from an unknown minimization principle (i.e. some kind of 'energy' is minimized through fractals).
- *Dynamical evolution.* Self-similarity can be related to a peculiar dynamic evolution of the system. This process is known as self-organized criticality (SOC) (Bak, Tang, and Wiesenfeld, 1987). Actually, SOC only explains how a system could develop a fractal shape. Since it does not clarify why the steady state is power law the question remains open. Nevertheless, this concept has had quite a success in statistical physics and biology and it is likely to play a role in the study of scale-free networks also. Very likely, this dynamical evolution can be driven by minimization principles and therefore this ingredient is closely related to the above one.
- *Multiplicative processes.* Because of the importance of the above questions, it is necessary to point out that sometimes the power laws *are not* a sign of something complicated and interesting going on in the system. Actually, in many cases the power laws arise from very simple processes. In nature we often find phenomena that are normally distributed. This is because the central limit theorem,[26] provides an excellent explanation for the ubiquity of normally distributed quantities. As an application of this theorem we find that in multiplicative processes (the value of a variable at time t is a percentage of its value at previous time t-1) we must expect variables either power law or log-normally distributed. Those two distributions have a similar shape and therefore in most cases, due to the small size, we can easily confuse them (see Section C.3.1).
- *Power laws from exponentials.* Finally, we can also obtain real power laws from exponential distributions if we introduce thresholds or sample the data in a particular way (Reed and Hughes, 2002).

---

[26] If we sum together different identically distributed stochastic variables we obtain a new variable that is normally distributed. This holds whatever the distribution of the original variables, provided their variance is finite.

Here we present the mechanisms and the models that produce power-laws. We start from the very traditional ones such as the random walk. We then present a brief exposition of minimization principles. The main part of the chapter is devoted to the discussion of multiplicative processes. This phenomenon either produces true power laws or log-normal distributions. Either way, this is one of the most important mechanisms in the formation of scale-free networks.

4.1   Random walk, Laplace equation, and fractals

Studies on the random walk originated with the experimental evidence about the motion of particles produced by the botanist Robert Brown (1828).[27] The simplest possible model (see, for example, Reichl 1980) is given by a particle starting in $y = 0$ and constrained on a one-dimensional line (it can move only by a step of length one to the left or to the right). If no particular preference is given, the particle moves to the left or the right with equal probability $p_L = p_R = 1/2$. Starting from the initial position (as shown in Fig. 4.2), the walker starts to wander around. The trajectories remain centred around the position $x = 0$ but the average distance from this position grows with time. The net displacement m (an integer) after N steps is given by composing all the moves n_L on the left (let us assume they are negative) and all the moves n_R to the right (let us assume they are positive). That is

$$n_R - n_L = m$$
$$n_R + n_L = N. \tag{4.98}$$

For a large series of steps the probability distribution for the displacement will be given by a Gaussian distribution

$$P(m) = \frac{1}{\sqrt{2\pi N}} e^{-\frac{m^2}{2N}}. \tag{4.99}$$

---

[27] Born in Montrose (Scotland) 1773 and died in London in 1858. His accurate observations revealed the problem of the motion of microscopic particles. The current explanation has been given in the framework of kinetic theory by A. Einstein.

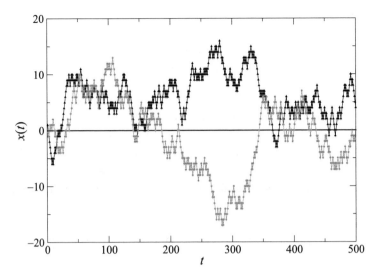

Fig. 4.2 Two different realizations of a one-dimensional random walk made of 500 steps of unitary length.

This is the first step in order to describe the property of Brownian motion. To improve this description we must pass to three dimensions, and use a step length l. For the latter point we have that if (on average) one step has length l the net displacement is $x = ml$. Now we also assume that the portion Δx of trajectory observed is much larger than the typical step size l (this is the case for Brownian motion where l is the intermolecular free path, while Δx can be of the order of millimetres).

☙ For a large number of steps the probability that the particle is in the interval $x, x + \Delta x$ is given by

$$P(x) = \frac{1}{\sqrt{2\pi N l^2}} e^{-\frac{x^2}{2Nl^2}}. \tag{4.100}$$

In three dimensions this formula does not change very much apart from the fact that x becomes a three-dimensional vector $\mathbf{x}$ (and apart from the normalization constant). Therefore, a particle moving by random walk is characterized by a Gaussian distribution of the displacement around the mean. Surprisingly a small modification of this process is one of the most studied fractal generators.

Fig. 4.3 A cluster grown according to the model of diffusion limited aggregation (DLA). For this specific case, more than $30,000$ different walkers started from the boundary of the system and deposited on the structure. The last $5,000$ particles have been drawn in lighter colour to show that time evolution of the object happens by growth on the tips. This means that the inner part is stable, i.e. it will not be filled by successive growth.

4.1.1   Diffusion limited aggregation

The above consideration of the microscopic dynamics of Brownian motion led to a cellular automaton called diffusion limited aggregation (DLA) (Witten and Sander, 1981). Cellular automata are similar to a board game like chess. You have the elements of the game (the chessboard, the pieces like the queen, the rooks, etc.) and rules of the game. In the DLA model the chessboard is made by a grid $L \times L$ (it is possible to relax this assumption and take any portion of space as chessboard). The pieces are particles that occupy the sites of this grid. In the middle of the grid we fix a particle ('seed'). This model of fractal growth is the prototype of most of the deposition and corrosion processes observed experimentally.

The game then starts following these rules:

- **Rule A: The chessboard.** A particle is put in the middle of the grid.
- **Rule B: Birth of a particle.** Another particle is added on the boundary of the grid. This particle starts a random walk as indicated in the following rule **C**.
- **Rule C: Life of a particle.** The particle moves from its starting node as a random walker. In particular, the particle has no memory of the

past path. At any point of the walk the particle can (with the same probability) move up, down, left or right.

- **Rule D: Deposition of a particle.** When the first walker passes near to the seed it sticks to it. This process is called deposition. After that, there are two deposited particles, and the next walker will deposit when it passes near one of the two. In the steady state, the various walkers deposited form an aggregate. After every deposition a new particle is added following rule **B**.
- **Rule D: End of game.** Stop the above process whenever the aggregate is comparable with the size of the grid.

With the rules defined above, we can understand the name of the model. We have a *diffusion* of the particles. This diffusion is *limited* by the boundary of the grid and by the *aggregation* of particles. For the first time in this chapter we note that by coupling thresholds and randomness in a dynamical process we can drive a self-similar behaviour. This idea will appear again and again and it is probably one of the main ingredients in the formation of self-similar objects. Despite the apparent simplicity of this model it gives rise to the very complicated object shown in Fig. 4.3.

A model of growth related to the diffusion limited aggregation is the dielectric breakdown model (DBM) (Niemeyer, Pietronero, and Wiesmann, 1984). The basic idea is to describe formation of natural structures like lightnings or artificial ones like the Lichtenberg figures.[28] These structures are formed by high voltage discharge inside an insulating material (dielectric). Insulating materials do not allow electric current to flow (i.e. electrons are bound to their atoms). If the applied electrostatic potential grows above a threshold, this field is strong enough to remove the electrons from the atoms; a discharge then takes place. An example of these figures is shown in Fig. 4.4 where a beam of accelerated electrons is shot at a transparent acrylic material. The physics of the process can be completely described by a simple statistical model that computes the electrostatic potential in the dielectric.

☢ This process is described by one of Maxwell's equations (Jackson, 1998) relating the divergence of the electric field **E** to the electrostatic charge ρ. This equation tells us that (in standard units of metre, kilogram and seconds, i.e. MKS)

$$\nabla \mathbf{E} = \frac{\rho}{\epsilon_0} \qquad (4.101)$$

---

[28] These are named after the German physicist Georg Christoph Lichtenberg (1742-99).

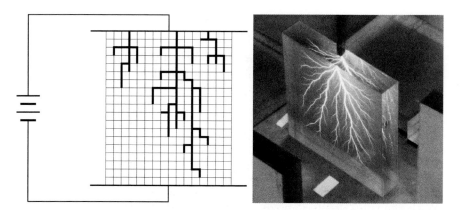

Fig. 4.4 Left: the set-up and result of dielectric breakdown model. Right: a Lichtenberg figure (courtesy of Bert Hickman, Stoneridge Engineering, http://www.teslamania.com, from Theodore Gray).

where $\nabla \mathbf{E} \equiv \frac{\partial \mathbf{E_x}}{\partial \mathbf{x}} + \frac{\partial \mathbf{E_y}}{\partial \mathbf{y}} + \frac{\partial \mathbf{E_z}}{\partial \mathbf{z}}$ and ϵ_0 is a parameter (the permittivity of free space). By using the definition of electrostatic potential, we can put this equation in the form of a Poisson equation

$$\nabla^2 \phi(x, y, z) = \frac{\rho}{\epsilon_0}. \tag{4.102}$$

In the case of a dielectric, the total charges in the medium are 0 so that the above equation becomes

$$\nabla^2 \phi(x, y, z) = 0. \tag{4.103}$$

The above expression is called Laplace equation. As regards the symbol ∇^2, this corresponds to computing and summing the second derivative of the function $\phi(x, y, z)$, that is

$$\nabla^2 \phi(x, y, z) \equiv \frac{\partial^2 \phi(x, y, z)}{\partial x^2} + \frac{\partial^2 \phi(x, y, z)}{\partial y^2} + \frac{\partial^2 \phi(x, y, z)}{\partial z^2}. \tag{4.104}$$

Let us now consider the medium as a regular square lattice, so that the physical quantities are defined only on the vertices of a grid. Let us now compute the value of the electrostatic potential on these points when an external electric field is applied. For every point with coordinates (x, y, z) it must be $\nabla^2 \phi(x, y, z) = 0$, but since the coordinates are now integers we have that the derivatives can be computed as finite

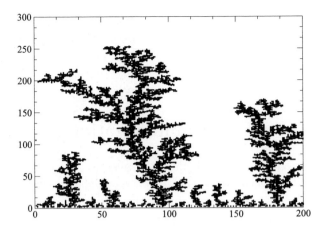

Fig. 4.5 A pattern of DBM grown in cylindrical geometry.

differences. That is

$$\frac{\partial \phi(x, y, z)}{\partial x} = \phi(x+1, y, z) - \phi(x, y, z) \qquad (4.105)$$

or alternatively

$$\frac{\partial \phi(x, y, z)}{\partial x} = \phi(x, y, z) - \phi(x-1, y, z). \qquad (4.106)$$

By first applying eqn 4.105 (*forward* finite difference) and then eqn 4.106 (*backward* finite difference) we obtain that

$$\frac{\partial^2 \phi(x, y, z)}{\partial x^2} = \frac{\partial}{\partial x} \frac{\partial \phi(x, y, z)}{\partial x} =$$
$$\phi(x+1, y, z) + \phi(x-1, y, z) - 2\phi(x, y, z). \quad (4.107)$$

Therefore the condition $\nabla^2 \phi(x, y, z) = 0$ in a three-dimensional simple cubic lattice now reads

$$\phi(x, y, z) = \frac{1}{6} \sum_{\xi, \eta, \zeta} \phi(\xi, \eta, \zeta), \qquad (4.108)$$

where (ξ, η, ζ) is one of the six neighbours of (x, y, z). That is, the value of the electrostatic potential at every point must be the average of the values in the neighbourhood. Using eqn 4.108 it is possible to compute recursively the value of the potential in the system.

We then model the breakdown stochastically. This means that we break the bonds with a probability proportional to their electrostatic field. When bonds are broken, the sites involved drop to zero potential. Therefore with this new boundary condition we need to recompute the values of ϕ on the sites left. The structure grows step by step, and the steady state of this process is presented in Fig. 4.5.

4.2   Power laws from minimization principles

One idea presented in order to explain the onset of self-similarity is the conjecture of *'feasible optimality'* (Marani *et al.*, 1998). Self-similar structures can arise to minimize some cost functions. This can be better explained through the following example: consider the problem of delivering water from the source to a series of different users. In this case a reasonable cost function is to require the least possible number of pipelines. This requirement can be complicated by other requests, such as to connect all the clients in such a way to have them as near as possible to the source (e.g. to reduce leakage probability and to improve the quality of the service). Different approaches are possible, as shown in Fig. 4.6

A star-like pipeline. Every vertex takes water directly from the central source. Personal convenience is maximized, but the total cost of pipelines is larger than the minimum one.

B chain pipeline, the total cost is minimized but personal advantage is completely lost. Only the vertices near the source have a good connection; those at the end of the chain are exposed to delays and increased probability of leakages.

C A self-similar solution.[29] A tree-like shape obtained by replicating the same structure is able to accomplish both requests.

This idea of fractal by minimization has been formulated more quantitatively in various forms by a number of authors. One version has been called feasible optimality (Marani *et al.* 1998; Caylor, Scanlon, and Rodríguez-Iturbe 2004) or minimization under constraint (Ferrer i Cancho and Solé, 2003). Often, this minimum configuration is only one of many different ones and therefore very difficult to reach during the evolution of the system. Fractals appear because self-similar configurations (with similar statistical properties corresponding to relative minima) occupy a relatively large region of configuration space. Therefore they are not only minimizing structures, but are also more easily accessible by system evolution. A specific example is

---

[29] Joking a little bit, we can consider A a capitalist solution, B a socialist one, and C in the spirit of Hegelian philosophers a 'synthesis' of capitalist-socialist struggle.

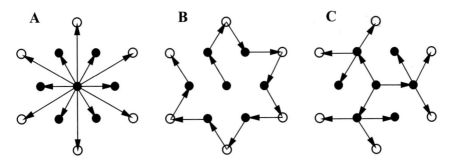

Fig. 4.6 Let us assume that this simple series of graphs represents a water pipeline starting from the source and running to the various houses. Pipes have a linear cost (also, a section cost which we shall not consider in this simple example). To save resources one would like to minimize the total pipe length. On the other hand any user wants to be as near as possible to the source; for example, to reduce failure risks or to have fresher water. Three classes of solutions are possible from the interplay of these two requests of minimization, one global and the other local. (A) Left: an egoistic approach. Everyone connects to the source. Maximum of local benefit (everyone is directly connected) but poor global optimization. If the edge of the triangular lattice measures 1 we need about 16.39 (i.e. $6(1 + \sqrt{3})$) pipeline units. (B) Centre: a minimum social cost approach. We need only 12 pipeline units, but distance from source for the unlucky last user can be as large as the system size. (C) Right: a self-similar structure that is a good compromise between A and B. Distance increases slowly and pipeline units needed are still 12.

that of river networks. As we shall see in Chapter 7, one hypothesis on river network evolution is that they sculpted the landscape in order to minimize the dissipation of total gravitational energy. Both experimental data and model results seem to confirm such a hypothesis.

4.2.1   Self-organized criticality

Closely related to this idea of evolution according to minimization there is the concept of *self-organized criticality* (Bak, Tang, and Wiesenfeld, 1987). In this process, the steady state of the evolution for a dissipative dynamical system is self-similar. The typical example of that is given by the class of models inspired by sandpiles. In the prototype of these models a series of sites can host different grains of sand until a certain threshold is reached. At the threshold, the site is said to become 'critical' and distributes the sand to its neighbours. It may happen that these can also become critical and this

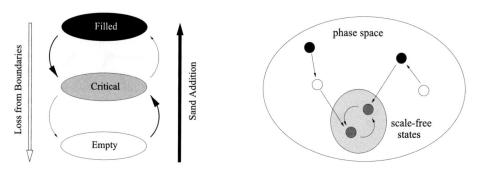

Fig. 4.7 The basic mechanism at the heart of self-organized criticality in sandpile models.

elementary process can create an avalanche on the large scale. 'Criticality'[30] (i.e. a power law distributed series of avalanches) is maintained by a feedback mechanism shown in Fig. 4.7 ensuring that whenever the system is full of sand no more grain can be added (since you have to wait for the activity to stop). Conversely, whenever the system is empty, no activity occurs, and many grains are added in a short time interval. As a result, the amount of sand tends to oscillate around a mean value. This is not a minimization mechanism, but rather a condition of dynamical equilibrium.

4.2.2   Entropy minimization

An example of this class is the distribution of words (Mandelbrot, 1953). Consider that in a language you have a series of n words. As evident from experimental data, we can assume that the frequency with which a word is used is given by a Zipf's law (see Section 3.6). The cost attached to the use of the j-th word is indicated by C_j. This cost of transmission is given by the letters used for the word, so that if d is the alphabet size, we can write $C_j = A \log_d(j)$, where A is a constant.

Suppose that language evolved in such a way, to minimize the unit transmission cost. The probability that a word j is used in a transmission is $p(j)$. The average cost per word can be defined as

$$C = \sum_{j=1,n} C_j p(j). \tag{4.109}$$

---

[30] The origin of the term critical traces back to the theory of critical phenomena. We remember here that around the critical point in a phase diagram all the physical quantities behave as power laws. Inspired by this fact, the conditions that generate power laws in these toy models have been called 'critical'.

In the case of linguistic systems the role of entropy is played by the information transmitted in a text. We can define the entropy H as

$$H = -\sum_{j=1,n} p(j)\log_2(p(j)). \tag{4.110}$$

Optimization of languages in this case means that the maximum amount of information is transmitted with the minimum length of the word. At least loosely, it makes sense to assume that languages evolved in order to transfer as much information as possible with the minimum cost. By imposing that the quantity C/H is the minimum, we can determine the form of the distribution of $p(j)$s.

 In formulas

$$min(C/H) \rightarrow \frac{d}{dp(j)}(C/H) = 0. \tag{4.111}$$

This value is given by

$$0 = \frac{d}{dp(j)}(C/H) = \frac{C_j H - C(\log_2(p(j)) + \log_2(e)\frac{p(j)}{p(j)})}{H^2}$$
$$= \frac{C_j H - C(\log_2(ep(j)))}{H^2} \tag{4.112}$$

where e is the base of natural logarithms. This brings us to the expression for the $p(j)$ given by

$$p(j) = d^{-HC_j/C}/e. \tag{4.113}$$

If we remember our previous assumption that $C_j \propto \log_d(j)$, we get a power law for the $p(j)$.

4.3   Multiplicative processes and normal distribution

While many power laws are originated by some 'complex' mechanism, some others have a very simple explanation. By using multiplicative processes we can obtain quite naturally both power laws and log-normal distributions (that can look like power laws) (Goldstein, Morris, and Yen, 2004). We do not enter here into the debate over whether observed data can be best fit by power law or log-normal variables. Here it is enough to note that the mechanism of multiplicative process is probably the most immediate model for fat-tail phenomena in nature since it naturally produces both.

Many textbooks and scientific papers deal with this topic. Some of them are very beautiful and complete and we suggest them for further reading (e.g. Mitzenmacher 2004; Newman 2005). Suppose you have an evolution process, where for example an organism transforms itself in time. As a general statement, the state S_t at time t will be determined by the previous states and by the external conditions. Whenever the state of the system can be written as

$$S_t = \epsilon_t S_{t-1}. \tag{4.114}$$

we have a multiplicative process. In other words, in a multiplicative process, the state of the system at time t is proportional to the state at time $t-1$. In biology this could represent the fact that the growth of an organism is ruled by its body mass at the previous step. In the case of city growth (Zanette and Manrubia, 1997; Manrubia and Zanette, 1998; Gabaix, 1999) this equation states that the population at a certain time step is proportional to what it was previously. In both cases the proportionality constant is given by the factor ϵ_t that can change its value at any time step. Turning back to eqn 4.114 we can immediately see that the variable S_t is determined by the product of the various ϵ_τ where τ is between 0 and t

$$S_t = \epsilon_t S_{t-1} = \epsilon_t \epsilon_{t-1} S_{t-2} = \epsilon_t \epsilon_{t-1} \epsilon_{t-2} ... \epsilon_2 \epsilon_1 S_0. \tag{4.115}$$

At this point the distribution probability of the state is related to that of the ϵs. Interestingly, regardless of the precise form of the distribution for the variables ϵ_τ the S_t is log-normally distributed.[31]

☕ To show that we can rewrite eqn 4.115 by taking the logarithms of both sides

$$\ln(S_t) = \ln(S_0) + \sum_{\tau=1,t} \ln(\epsilon_\tau). \tag{4.116}$$

In this way the product of the ϵs transforms in the sum of the logarithms.

This sum of the logarithms of the ϵ_τ (under very mild conditions) is a variable following a normal distribution (regardless of the distribution of the ϵ_τ). This result comes from the application of the 'central limit theorem'. This theorem states that, in certain very general hypotheses, the sum of identically distributed random variables with finite variance is a new stochastic

---

[31] We recall that the shape of log-normal function is given by

$$f(k) = P_{LN}(k) = \frac{1}{\sqrt{2\pi}\sigma k} e^{-\frac{(\ln(k)-\mu)^2}{2\sigma^2}}.$$

variable normally distributed. Therefore, if $\ln(S_t)$ is normally distributed, the variable S_t is log-normally distributed.

This very simple mechanism has been rediscovered and explained over and over many times since the definition of log-normal distributions in 1879 (McAlister, 1879). One of the first applications of these ideas traces back at least to the economist Gibrat (1930, 1931) who uses essentially this model under the name of proportionate effect. Using a different terminology, a somewhat similar idea was introduced at the beginning of last century for biological problems (Kapteyn 1903, 1918).

With this idea we have two possible outcomes. As explained above we have true log-normal distributions that can easily be confused with power laws. This happens whenever the log-normal distribution is studied in a range of k for which $\sigma >> \ln(k)$. On the other hand, a very similar situation also triggers the formation of true power laws as shown in the next subsection.

4.3.1   Power laws from multiplicative processes

A tiny modification of the above mechanism, namely the introduction of a threshold, has a dramatic effect on the results. In particular, *when multiplicative processes are coupled with fixed thresholds we have production of power-law distributions*. In the simple model of the economy we use for the demonstration (Champernowne, 1953), we study the evolution of a society where the various individuals are divided into classes according to their income. The minimum income possible is indicated by m. This means that the poorest people in this artificial society are in class 1 and have an income between m and γm. Going to the upper class (number 2) we have people with income between γm and $\gamma^2 m$. More generally a class j will rank individuals whose income is between $\gamma^{j-1} m$ and $\gamma^j m$. In this economy we assume that individuals change their class with a probability that depends only upon the distance between the classes. This means that the possibility of passing from class 1 (poorest) to class 3 is the same as passing from class 3 to 5. This can be described by a multiplicative process in which we move from one class to a higher one when the income increases by a factor γ (or powers of it). The threshold is in the basal class where people cannot have an income lower than m.

We assume that:

- there is no limit to the income value so that there is no maximum class;
- we change class at any time step;
- we assume $\gamma = 2$;
- we can change only to the next upper or next lower class with probabilities $p_- = p_{i,i-1} = 2/3$ and $p_+ = p_{i,i+1} = 1/3$;

- when considering class 1 $p_- = 2/3$ gives the probability of staying in this lowest class.

With the above rules every change of class corresponds to doubling or halving the income. The values of the probabilities have been carefully chosen in order to have a constant expectation value $E[x^t]$ for the income x^t. Indeed, if the income is x^t at time t in the next time step $t + 1$ it will increase to $2x$ with probability $1/3$ and decrease to $x/2$ with probability $2/3$ and therefore

$$E[x^{t+1}] = \frac{2}{3}\frac{x^t}{2} + \frac{1}{3}2x^t = x^t. \tag{4.117}$$

The probability of being in class j is simply $1/2^j$.

This can be seen by considering what happens to the population of the first two classes N_1^t, N_2^t. The equation rates are

$$N_1^{t+1} = \frac{2}{3}N_1^t + \frac{2}{3}N_2^t$$

$$N_2^{t+1} = \frac{1}{3}N_1^t + \frac{2}{3}N_3^t. \tag{4.118}$$

At the stationary state $N_i^{t+1} = N_i^t = N_i$ and therefore the first equation gives $N_1 = 2N_2$. Putting this value in the second equation, we find $N_2 = 2N_3$. All the other populations can be computed similarly, so that in general

$$N_{i+1} = 1/2N_i. \tag{4.119}$$

Since the population halves by increasing the class, the probability must also decrease accordingly. Therefore the probability of being in class j is $1/2$ of the probability to be in the class $j - 1$. This together with the requirement that the sum of all the probabilities is equal to 1 gives the above expression $1/2^j$.

Henceforth the probability to be in a class larger than (or equal to) j is $1/2^{j-1}$.

This is a consequence of the fact that

$$\sum_{k=j,\infty} \frac{1}{2^k} = \frac{1}{2^{j-1}}. \tag{4.120}$$

All the above results allows us to write

$$P(X > 2^{j-1}m) = 1/2^{j-1} \rightarrow P(X \geq x) = m/x. \tag{4.121}$$

This multiplicative process therefore results in a power-law distribution. Note that in this way we recover Pareto's law which states that if we rank the people in a society according to their income we obtain a power-law distribution.

4.3.2   Combination of exponentials

We have already seen that the (Gaussian) normal distribution is widespread in various phenomena. Since in some cases the power laws arise from normal variables, the ubiquity of the latter ones can be an explanation for the observed scale invariance.

Consider a process that grows exponentially and consider that we can observe it only at certain random times (Reed and Hughes, 2002). The latter situation can be applied to earthquakes where systematic recording of the data began only recently. The distribution of these observed states behaves as a power law, even if the general distribution does not.

Let us see the simplest example of an exponential process of growth where a variable (as example the size of a population in an environment with infinite resources) grows exponentially with time $x(t) = e^{\mu t}$. Individuals (or whatever the x represents) also have an extinction probability given by $P(t) = e^{-\nu t}$ where $\nu > 1$. The distribution of the state $X = e^{\mu T}$ (where T is extracted from the previous $P(t)$) is a power law:

$$F_X(x) = \left(\frac{\nu}{\mu}\right) x^{-\frac{\nu}{\mu}-1}. \tag{4.122}$$

This is only a particular case where two exponentials concur to form a power law. Let us see the general case (Newman, 2005). If the quantity y is exponentially distributed

$$P_y(y) = e^{\alpha y} \tag{4.123}$$

and we are interested in another quantity x given by

$$x = e^{\mu y} \rightarrow y = \frac{1}{\mu} \ln x \tag{4.124}$$

we find that this latter quantity x is distributed according the power law function shown in eqn 4.122.

Let us see why: a basic property of distribution of probabilities is that the probability is a measurable quantity and it must be recovered regardless the distribution used. From conservation of probability $(P_x(x)dx = P_y(y)dy)$ it follows that the distribution of xs is power-law.

☕ In the above case, from

$$P_x(x)dx = P_y(y)dy \tag{4.125}$$

it follows that

$$P_x(x) = P_y(y)\frac{dy}{dx} = \frac{e^{\alpha y}}{\mu x} = \frac{x^{\alpha/\mu}}{\mu x} = \frac{1}{\mu}x^{\alpha/\mu - 1}, \qquad (4.126)$$

which is exactly eqn 4.122 where $\alpha = -\nu$.

4.4   Preferential attachment, the Matthew effect

One of the most successful applications of multiplicative processes is given by preferential attachment. To date, this is the most successful mechanism adopted in the study of growing networks. Interestingly, the idea that we are going to explain has been independently rediscovered several times in different fields and ages. Precisely for this reason it has also been given several names. For example: *Yule Process, Matthew effect, Rich gets richer, Preferential Attachment, Cumulative advantage*.

In the community there is some agreement (Mitzenmacher, 2004; Newman, 2005) that the first to present this idea has been G. Yule (1925) in order to explain the relative abundance of species and genera in biological taxonomic trees. As shown in Chapter 8 when considering a set of biological species we have that the classification (taxonomic) tree has scale-free properties. The null hypothesis consists in considering that the set of species arises from a common evolution. Therefore we consider one parent species and after mutation we obtain a new one that very likely can be grouped in the same genus. Every now and then though, speciated species (the new ones) can be so different from the parent one that they can form a new genus on their own (or be grouped in an existing different one). The probability of speciating will be larger for genera that are already large, since mutation rate is constant for any individual.

This explanation allow us to focus on the two ingredients of the model. Firstly you have to invoke a certain a priori dynamics (hereafter called **growth**). Secondly, this dynamics selects successful elements and makes them even more successful (hereafter called **preferential attachment**). In detail, take a set of elements each of which is characterized by a certain number N_i^t. As a possible example this could be the number of different genera that have i species per genera. The set can also be a set of vertices in a graph and the number N_i^t can represent the number of pages whose in-degree is i. Now let us introduce a rule that introduces new elements in the set; these elements will not be shared equally between the older ones, but rather will be assigned more to those that already have many. Let us consider that N_i^t gives the number of vertices with certain degree i (the total number of vertices is t). The probability p_+ that the number N_i^t increases by edge addition is given by two terms:

1. a random selection of the vertices origin and destination for the edges;

2. a choice of the vertices proportional to their degree.

A similar form holds also for the probability p_- that this number N_i^t might decrease (this happens when vertices whose degree is i become vertices with degree $i+1$).

 This can be written as (Mitzenmacher, 2004)

$$p_+ = \alpha N_{i-1}^t/t + (1-\alpha)(i-1)N_{i-1}^t/t$$
$$p_- = \alpha N_i^t/t + (1-\alpha)(i)N_i^t/t. \tag{4.127}$$

So that the balance equation can be written as

$$\frac{dN_i^t}{dt} = p_+ - p_- = \frac{\alpha(N_{i-1}^t - N_i^t) + (1-\alpha)((i-1)N_{i-1}^t - iN_i^t)}{t}. \tag{4.128}$$

At this point we make the crucial hypothesis that in the steady state the N_i^t are linear in time, that is $N_i^t = c_i t$. The result of the computation is that the coefficients c_i (and then the degrees at fixed time t) are power-law distributed

$$c_i \propto i^{-\frac{2-\alpha}{1-\alpha}}. \tag{4.129}$$

Starting from the first one, N_0^t, for which $c_0 = N_0^t/t$, we have

$$\frac{dN_0^t}{dt} = c_0 = 1 - \frac{\alpha N_0^t}{t} = 1 - \alpha c_0 \tag{4.130}$$

we obtain the first value $c_0 = 1/(1+\alpha)$. The next values are obtained through a recurrence equation

$$c_i(1 + \alpha + i(1-\alpha)) = c_{i-1}(\alpha + (i-1)(1-\alpha)). \tag{4.131}$$

For large values we have

$$\frac{c_i}{c_{i-1}} = 1 - \frac{2-\alpha}{1+\alpha+i(1-\alpha)} \simeq 1 - \left(\frac{2-\alpha}{1-\alpha}\right)\left(\frac{1}{i}\right). \tag{4.132}$$

We do not explicitly write the derivation of the result, but note that if we assume for the tail of the distribution (i.e. large degrees)

$$c_i \propto i^{-\frac{2-\alpha}{1-\alpha}}, \tag{4.133}$$

we find exactly the previous result

$$\frac{c_i}{c_{i-1}} = \left(\frac{i-1}{i}\right)^{\frac{2-\alpha}{1-\alpha}} \simeq 1 - \left(\frac{2-\alpha}{1-\alpha}\right)\left(\frac{1}{i}\right). \tag{4.134}$$

Along these lines Simon (1955, 1960) proposed a model not only for species abundance, but for a variety of different situations (population in the cities, words in documents, papers published, and incomes in a society). The same mechanism then became familiar also in the field of economics under the name of Simon's model. R.K. Merton (1968, 1988) formulated a similar idea (even if not at that level of mathematical formulation) in the field of scientific production (Zuckerman, 1977). The same author associated this principle to sentence from Matthew's Gospel (New American Bible, 2002) where it is said: *For to everyone who has, more will be given and he will grow rich [...]*, (Matthew 25:29; New American Bible). After that, this phenomenon was also known as 'Matthew Effect'. The idea of this work can be summarized by a quotation by a Nobel laureate: 'The world is peculiar in this matter of how it gives credit. It tends to give the credit to the [already] famous people' (Merton, 1968). Interestingly, exactly this problem of the study of formation of scientific consensus is at the heart of the work of D. J. de Solla Price (1965) who a few years before Merton published another version of this 'rich gets richer' model. The latest and most interesting formulation of this concept is explained in great detail in Chapter 5 where the Barabási-Albert model (Barabási and Albert, 1999) is presented and analysed. In this latter case the focus is on graphs. Vertices are the elements of the system and measure of their success is their degree.

5.   Graph generating models

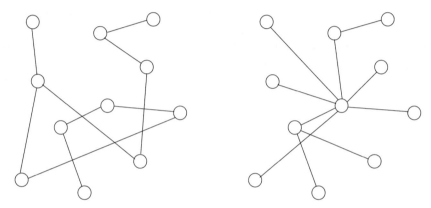

Fig. 5.1 Left: realization of an Erdős-Rényi Graph. Right: a realization of a Barabási-Albert model. The number of vertices and that of the edges is equal in both cases.

Christmas won't be Christmas without any presents and science won't be science without any models. The scientific approach is based on the idea that phenomena can be described quantitatively by means of a schematic, simplified representation usually called 'model'.

The very idea of a graph is already a powerful simplification for different phenomena. Graph models are introduced in order to describe how these structures originate and how they can evolve (see Fig. 5.1). Models (as cellular automata) are defined by giving a set of rules. Ideally a network model must produce graphs whose properties reproduce those of the real data. If this is not the case, the rules of the model need to be changed. By using a model we can reduce the complexity of the real world to a stage that we can understand and get used to. This line of proceeding was introduced long before the beginning of modern science. Presocratic Greek philosophers like Thales, Anaximander, and Heraclitus in their quest for the ἀρχή (the first principle) represent an ancient tendency to reduce the whole universe to a simple representation.[32] By defining a set of mathematical quantities measurable in the models and in the data one can test the validity of a model representation. Sometimes models work remarkably well against any expectation. One of the major successes of Newtonian mechanics (the study of motion) starts from the very unrealistic approximation of no friction. Even so, we can predict with great accuracy the time evolution of a body in motion.

In the field of networks there has been a continuous feedback between data properties and mathematical abstraction. This is at the foundation of the activity in this field and has produced a variety of different models that we are going to present. For all the models listed in this chapter we give at least the information on degree distribution, diameter distribution, and clustering. One must remember that not all of them (even if reproducing most of the quantitative behaviour) are equally likely. In general, models are important to show only some aspects of the phenomenon but they might be rather distant from a real application. Let us clarify this point with an example. We present in the last chapter of the book the experimental evidence that even a sexual intercourse network is characterized by power-law distribution. In this system the vertices are people and an edge is drawn if sexual intercourse occurs between two of them. If we were to use the *random graph* model we could say that to have sexual intercourse between two people, one of them tosses a coin for every possible partner (equally likely) and the other must accept the proposal. In a *'small world'* each one would have sexual intercourse with relatives and acquaintances; very seldom with someone new. In a *copying model*, one would have sexual relations by borrowing the partners of the best friends. In the *Barabási-Albert* model, any new individual entering a circle of friend establishes a new intercourse. To choose with whom, the person must know the past history of the new friends. The larger the number of past acts of intercourse, the better. Finally in a *fitness model* one would have intercourse if both persons are beautiful (i.e. their fitnesses are above a certain threshold). Note that while the last model may seem more reasonable, there are problems with it too. Indeed the only way for an ugly person (low fitness) to have sex would be to find a partner with extraordinary beauty (large fitness).

---

[32] Anyway, a mathematical description was missing, therefore the possibility of predicting the future behaviour of a system quantitatively was impossible.

This chapter is devoted specifically to graph models. Some of them produce scale-free networks some do not. In general through modelling we can understand the properties of real data. The first two models presented, namely the Random Graph model and the Small World model, do not form scale-free networks. We then present the Barabási-Albert model and a series of other models such as the fitness and the copying models, some of which do produce scale-free degree distributions.

5.1 Random graph model

The simplest model that has been introduced is due to the two mathematicians Paul Erdős[33] and Alfréd Rényi[34] (Erdős and Rényi 1959, 1960, 1961). A similar idea (in the completely different context of polymer physics) was previously introduced to explain the phenomenon of percolation (Flory, 1953), here we follow the traditional random graph exposition.

A real situation inspiring the model could be that of the telephone wiring of a set of houses. The vertices (fixed) are a parameter of the model, the edges (i.e. telephone cables) can be drawn or not, and distances in a first approximation can be neglected (all the edges are equal for short distances). In this spirit the first rule is to take a fixed amount n of vertices in the graph. After that we try to determine a rule in order to decide how many edges (cables) must be drawn. The simplest choice is to say that all the edges have the same probability of existing. This corresponds to sampling every one of the possible $n(n-1)/2$ edges and drawing the edge with a certain fixed probability p. There is another definition for the random graph model. We take all the possible realizations of a graph with n vertices and m edges. Unless otherwise specified in the following we mean as a random graph a graph obtained according to the first method.

In Fig. 5.2 we present two different realizations of a random graph. They look different but their statistical properties are the same. Since the extraordinary amount of efforts in this field, many quantities have been computed

---

[33] Paul Erdős was born in Budapest in 1913 and died in Warsaw in 1996. He was one of the most productive mathematicians of his era, and was generally regarded by colleagues as one of the most brilliant minds in his field.

[34] Hungarian mathematician, born in Budapest in 1921 and died in the same place in 1970. He was the founder and director of the Mathematical Institute of the Hungarian Academy of Science.

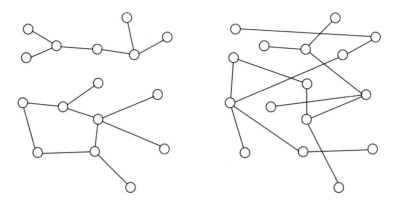

Fig. 5.2 Two different realization of a Random Graph both with $n = 16$ and $p = 0.125$.

analytically. Here we present only the very simple ones without devoting too much space to formal derivation and proofs.

- A first trivial quantity that can be computed is the expected value of the **size** of the graph. This value gives the total number of edges present at the end of the graph construction. Since the probability is p and the total number of trials is given by the $n(n-1)/2$ possible edges, we have that the expectation value is given by the product of the two.

$$E(n) = p\frac{n(n-1)}{2}. \qquad (5.135)$$

Consequently, the probability of having at the end of the process a graph $G(n, m)$ is

$$P(G(n, m)) = p^m (1-p)^{\frac{n(n-1)}{2} - m}. \qquad (5.136)$$

Following a similar derivation, the **average degree value** is twice the number of edges divided by the number of vertices. The factor two is because the same edge is counted twice: in the degree of the origin vertex and in that of the destination vertex.

$$\langle k \rangle = 2m/n = p(n-1) \simeq pn. \qquad (5.137)$$

- We can also compute the **degree probability distribution**. Let us start with the hypothesis that the number of vertices is so large that we can forget about the correlation between the various degrees in the graph (the various degrees are not independent of each other in a finite graph). To obtain a vertex whose degree is k, we must have k times a successful event whose probability is p and $(n-1-k)$ times an unsuccessful event whose probability is $(1-p)$.

 Since this can happen in $\binom{n-1}{k} = \frac{(n-1)!}{(n-1-k)!k!}$ combinations we have

$$P_k = \binom{n-1}{k} p^k (1-p)^{n-1-k} = \frac{(n-1)!}{(n-1-k)!k!} p^k (1-p)^{n-1-k}.$$

(5.138)

This distribution is automatically normalized since

$$\sum_{k=1,n-1} P_k = (p + (1-p))^{n-1} = 1.$$

(5.139)

The above distribution is called a *binomial distribution* and describes processes where you have two distinct possible mutually exclusive outcomes (in this case either you draw an edge or not). We remember that k is an integer, and therefore the degree distribution is a discrete one. It is easy to see that a binomial distribution is not self-similar.

 This distribution is usually approximated by means of the Poisson distribution in the two limits $n \to \infty$ and $p \to 0$ (when np is kept constant and $n-1 \simeq n$) we have (Gnedenko, 1962):

$$P_k = \frac{n!}{(n-k)!k!} p^k (1-p)^{n-k} \simeq \frac{(np)^k e^{-pn}}{k!}.$$

(5.140)

Since the mean value $\langle k \rangle$ of the above distribution is given by np we can write

$$P_k = \frac{\langle k \rangle^k e^{-\langle k \rangle}}{k!}.$$

(5.141)

- We can give an estimate, even if not accurate, for the **clustering coefficient**. The probability that two neighbours of a vertex are also neighbours of each other (thereby forming a triangle) is still p apart from some correcting terms (we are neglecting the fact that they have a common neighbour).
- Similar reasoning allows us to estimate the **diameter**. Consider the number N_i^1 of first neighbours of a vertex i. This number is the degree k_i of i whose average is $\langle k \rangle$. Therefore we have

$$N_i^1 = \langle k \rangle = pn.$$

(5.142)

Now consider the number N_i^2 of the vertices that are two edges apart from i. We make an approximation: we assume that this set is composed by the neighbours of the first neighbour. In general that is false, because a neighbour of the first neighbour can be a first neighbour itself (or the starting vertex i). In this approximation

$$N_i^2 \simeq N_i^1 \langle k \rangle = \langle k \rangle^2. \tag{5.143}$$

This approach is analogous to the exploration of the graph made with breadth-first search algorithm (see Section 1.2). In general, through this approximation (which becomes more and more crude as the distance from i grows) we can say that for any distance d the number of the vertices that are d-th neighbours grows as $(\langle k \rangle)^d$. In the worst hypothesis this procedure must end at least when d is equal to the diameter D (it can stop well before if the starting vertex is central in the graph). At distance D we therefore have

$$N_i^D \simeq \langle k \rangle^D. \tag{5.144}$$

The total number of vertices n is given by

$$n = N_i^D + N_i^{D-1} + ... + N_i^1, \tag{5.145}$$

which at the leading term can be approximated by

$$n \simeq N_i^D \simeq \langle k \rangle^D. \tag{5.146}$$

Taking the logarithm of the above formula we obtain that

$$D \simeq \frac{\ln(n)}{\ln(\langle k \rangle)}. \tag{5.147}$$

That is, the diameter D grows as the logarithm of the size of a random graph. Therefore if the number of vertices increases from $10,000$ to $100,000$, the diameter simply increases from about 4 to about 5. In the case of sparse random graphs, under the hypothesis that $np \to \infty$ it has been shown analytically (Chung and Lu, 2001) that 'almost surely' the diameter D varies as $(1 + o(1)) \frac{\ln(n)}{\ln(np)}$.

5.2   The small-world model

The small-world model (Watts and Strogatz, 1998; Watts, 1999) explains why the diameter of real graphs can remain very small when the number of vertices increases (small-world effect).[35]

---

[35] We have seen in the previous section that in the random graph model the diameter increases logarithmically with respect to the number of vertices. This is a common feature in most if not all graph models.

In this model one starts with a portion of an ordered grid. In a grid any vertex occupies a precise position in the space and therefore it is possible to define a Euclidean distance between them. The shortest possible distance is the grid unit. Vertices apart only one grid unit are connected (upper part of Fig. 5.3); they form the set of the first neighbours. This local connectivity can be increased by adding new edges to the nearby vertices, as shown in the middle part of Fig. 5.3. The small-world model is based on the idea that other random connections can also be established between random vertices; either by rewiring existing edges (original formulation) or by adding brand new ones (more recent formulation) with probability p. Both methods produce similar structures. At this point we have a small-world graph. In the lower part of Fig. 5.3 we show two of them, originated by a one-dimensional and a two-dimensional regular grid respectively.

In this model, there are two parameters one can tune

1. The *coordination number* z that gives the number of vertices directly connected in the regular structure. In a one-dimensional ($d = 1$) system with $j = 3$ (connections arrive to the third layer) every vertex has connections with $z = 6$ other vertices (three from one side and three from the other). This number of connections also grows with the dimensionality. In general we can write

$$z = 2jd. \qquad (5.148)$$

   In the example shown in Fig. 5.3, j is 2 and therefore $z = 4d$. That is, $z = 4$ for the one-dimensional system and $z = 8$ for the two-dimensional system. If the initial number of vertices (order) is n, the size (i.e. the number of edges) is $m = nz/2$.

2. The *shortcuts probability* p that gives the probability per existing edge to draw a new edge (shortcut) between two random vertices. This means that the total number of shortcuts is

$$mp = nzp/2. \qquad (5.149)$$

   To remove the 2 in this formula, we follow the convention (Newman, Moore, and Watts, 2000) for which the coordination number is defined as $z' = z/2$. In this way the total number of shortcuts becomes $nz'p$.

As regards the quantities of this model we assume to consider the situation where p is so small that essentially we are considering a regular grid. In the opposite limit of all shortcuts (obtained by rewiring), the underlying structure disappears and we obtain a random graph.

- We have that the **degree distribution** is a function peaked around the fixed value z characteristic of the regular grid. With no shortcuts,

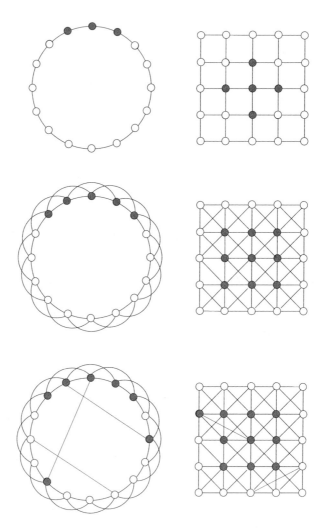

Fig. 5.3 On the first row a regular one-dimensional lattice on the left and a regular two-dimensional lattice on the right. In the second row the same lattices with extra edges increasing the local connectivity. On the last line we have the small-world lattices with shortcuts.

the distribution is not even a regular function, but it is zero elsewhere apart from z (it is a delta function different from zero only in z and zero otherwise). When shortcuts are many and there is no more underlying grid we must expect a behaviour similar to that of random graph. That is, in this limit we have a Poisson distribution. For the original version of the model (with rewiring) the general form of the distribution is rather complicated. We report here the results for the version with added edges (Newman, 2003). In case some shortcuts are present (and old edges have not been rewired) every vertex has at least the $2z$ edges of the regular structure. If we neglect the correlation between the vertices in the large order limit ($n \to \infty$) we can say that this distribution is still a binomial one.

 Therefore we find that the probability of having a degree k is

$$P_k = \binom{n}{k-2z} \left(\frac{2zp}{n}\right)^{k-2z} \left(1 - \frac{2zp}{n}\right)^{n-k+2z} \tag{5.150}$$

for $j \geq 2z$ and 0 otherwise. It is easy to realize that also for intermediate values of p the small-world model does not produce scale-free networks.

- Precisely the shortcut presence is the 'active ingredient' at the heart of the small-world effect. Even if these shortcuts are very few, their effect is dramatic (Amaral and Barthélemy, 1999). Using numerical simulation we can compute the variation on the **diameter**. It has been shown that a system with $N = 1,000$ vertices ($d = 1$), a coordination number $z = 10$, and a rewiring probability $p = 1/4 = 0.25$ has a diameter as small as $d = 3.6$. With no rewiring at all, the diameter of the same system is $d = 50$. Even with p as small as $p = 1/64 = 0.015625$ one still finds a small diameter $d = 7.6$. The hypothesis that the appearance of small-world behaviour is not an abrupt phenomenon, but rather a crossover has been tested (Barthélemy and Amaral, 1999a). It has been found that is the crossover size over which the network behaves as a small world is given by (Barthélemy and Amaral, 1999b) $N^* \propto p^{-\tau}$ with $\tau \simeq 1$.

- Newman, Moore, and Watts (2000) propose an analytical expression for the mean distance l

$$l = \frac{n}{z'} f(npz') \tag{5.151}$$

where $z' = z/2$ and the function $f(x)$ is

$$f(x) = \frac{1}{2\sqrt{x^2 + 2x}} \tanh^{-1} \frac{x}{\sqrt{x^2 + 2x}}. \tag{5.152}$$

- The **clustering coefficient** of the whole network is usually very high and it is reminiscent of the regular connection of the underlying grid. As long as z stays reasonably small and in particular $z < \frac{2}{3}n$ (as is the case when $n \to \infty$), we have:

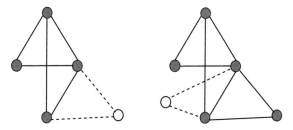

Fig. 5.4 Two steps in the construction of a growing network with preferential attachment.

 For the original formulation (with rewiring) (Barrat and Weigt, 2000) we have

$$C = \frac{3(z-1)}{2(2z-1)}(1-p)^3 \qquad (5.153)$$

while for the formulation without rewiring (Newman, 2002b), it can be shown that

$$C = \frac{3(z-1)}{2(2z-1)+4zp(p+2)}. \qquad (5.154)$$

5.3   The Barabási-Albert model

The Barabási-Albert model (Barabási and Albert, 1999; Albert, Jeong, and Barabási, 2000) is specifically suited to reproduce the striking evidence of time growth of many real networks (e.g. Internet and WWW). To reproduce this feature the graph is built through successive time-steps when new vertices are added to the system. Also the number of edges increases time, since the new vertices connect to the old ones. The vertices destination (those already present) are chosen with a probability that is proportional to their degree at the moment.

In Fig. 5.4 we present two steps in this construction. Schematically, we can say that the two ingredients of the model are *growth* and *preferential attachment*. Growth implies that new vertices enter the network at some rate. Preferential attachment means that these newcomers establish their connections preferentially with vertices that already have a large degree (*rich-get-richer*). This latter rule is in the spirit of the Matthew effect described in Section 4.4. More quantitatively, this model can be reconnected to a Yule process (Yule, 1925) described in the same section. Growth and preferential attachment are specifically suited to model the Internet and the World Wide Web (though the latter is directed while the former is not), two networks that in a relatively short timespan (fifteen to twenty years) have seen a huge growth of their elements. Furthermore, new routers (for the Internet) and new web pages (for the WWW) tend to connect with authoritative pre-existing routers and web pages, where authoritativeness is based on the consensus and can be weighted by the number of connections.

The steps of the graph generating algorithm are as follows

1. We start with a disconnected set of n_0 vertices (no edges are present).

2. New vertices enter the system at any time step. For any new vertex m_0 new edges are drawn.

3. The m_0 new edges connect the newcomers' vertices with the old ones. The latter are extracted with a probability $\Pi(k_i)$ proportional to their degree, that is

$$\Pi(k_i) = \frac{k_i}{\sum_{j=1,n} k_j}. \tag{5.155}$$

Note that, since at every time step only one vertex enters, we have for the order and the size of the network respectively

$$n = n_0 + t$$

$$m = 1/2 \sum_{j=1,n} k_i = m_0 t. \tag{5.156}$$

These two simple rules produce naturally scale-free networks in the sense that the degree distribution is power-law distributed $P(k) \propto k^{-\gamma}$.

A rigorous derivation for the analytic form of $P(k)$ is possible but here we present the simplest original derivation. Here we consider the degree as a continuous variable. New vertices enter the network at a constant rate. At time t the old ones are $n = n_0 + t - 1$. The first quantity we can derive is the variation of the degree with time.

$$\frac{\partial k_i}{\partial t} = A\Pi(k) = A \frac{k_i}{\sum_{j=1,n} k_j} = \frac{Ak_i}{2m_0 t}. \tag{5.157}$$

The constant A is the change of connectivity in one time step, therefore $A = m_0$. Since at initial time t_i the initial degree is $k(t_i) = m_0$ we have

$$\frac{\partial k_i}{\partial t} = \frac{k_i}{2t} \rightarrow k_i(t) = m_0 \left(\frac{t}{t_i}\right)^{1/2}. \tag{5.158}$$

This simple computation shows that in a Barabási-Albert model the degree grows with the square root of time. This relation allows us to compute the exponent of the degree distribution:

The probability $P(k_i < k)$ that a vertex has a degree lower than k is $P(k_i < k) = P(t_i > \frac{m_0^2 t}{k^2})$. Since vertices enter at a constant rate, their distribution is uniform in time, that is $P(t) = A$, where A is a constant. The value of A can be determined by imposing normalization of the distribution. This means $\int_0^n A = 1$, which gives $A = P(t) = 1/n = 1/(n_0 + t)$. In this way, we can write

$$P(t_i > \frac{m_0^2 t}{k^2}) = 1 - P(t_i \le \frac{m_0^2 t}{k^2}) = 1 - \frac{m_0^2 t}{k^2} \frac{1}{(n_0 + t)} \qquad (5.159)$$

from which we have

$$P(k) = \frac{\partial P(k_i > k)}{\partial k} = \frac{2m_0^2 t}{(n_0 + t)} \frac{1}{k^3}. \qquad (5.160)$$

Therefore, we find that the degree distribution is a power law with a value of the exponent $\gamma = 3$.

For this model some results have been obtained:

- The **degree distribution** is scale invariant only if the preferential attachment rule is perfectly linear; otherwise the degree is distributed according to a stretched exponential function.
- As regards the **diameter** D of Barabási-Albert networks, an analytical computation (Bollobás and Riordan, 2004) shows that $D \propto \ln(n)/\ln(\ln(n))$.
- The **clustering coefficient** of a Barabási-Albert model is five times larger than those of a random graph with comparable size and order. It decreases with the network order (number of vertices). Some analytical results are available in the particular limit of large and dense graphs (Fronczak, Fronczak, and Holyst, 2003; Barrat and Pastor-Satorras, 2005).

5.4   Modifications to the Barabási-Albert model

5.4.1   Fitness and preferential attachment model

One immediate generalization of the above model (see also Section 5.6) is given by allowing the various vertices to have a specific feature. This individual feature called fitness (Bianconi and Barabási, 2001*b*) introduces a novel behaviour in the Barabási-Albert model. The rules of this model are the same as the original one:

1. We start with n_0 different vertices characterized by a constant ability (*fitness*) η_i to attract new edges. The η_i are extracted from a probability distribution $\rho(\eta)$.

2. The growth remains with new vertices entering the system with their new fitnesses η_i.

3. The preferential attachment is slightly modified, taking into account the fitnesses. The edges are drawn towards the old vertices with a probability $\Pi(k_i, \eta_i)$

$$\Pi(k_i, \eta_i) = \frac{\eta_i k_i}{\sum_{j=1,n} \eta_j k_j}. \tag{5.161}$$

The presence of fitness in this case mitigates the rich gets richer effects. New vertices added lately to the system may increase their degree and may become the most successful.

- We can derive analytically the form of the degree distribution that is now dependent upon the form of the fitness distribution $\rho(\eta)$. In the case of a uniform distribution (i.e. $\rho(\eta)$ constant), we have that (Bianconi and Barabási, 2001b)

$$P(k) \propto \frac{k^{-(1+C^*)}}{\ln(k)} \tag{5.162}$$

  where $C^* = 1.255$ is a constant whose value is determined numerically.
- While no particular result is known for the clustering, this model develops non-trivial disassortative properties that make it a very good model to reproduce Internet autonomous systems properties (Vázquez, Pastor-Satorras, and Vespignani, 2002).

It is interesting to note that in particular conditions this network can develop a transition where (even in the limit of infinite size) a single vertex can have a finite fraction of all the edges in the system (Bianconi and Barabási, 2001a). An extension of this model where only the fitness is present and without growth and preferential attachment produces scale-free networks in certain conditions (see Section 5.6).

5.4.2   Edges growth

Not only the vertices but also the edges can 'grow'. In particular, we can allow new edges to be added between existing vertices (Krapivski, Rodgers, and Redner, 2001). The motivation of this model was to provide a more realistic model for study of the World Wide Web. Indeed in this specific network (as turned out to be the case also for Wikipedia) most of the modifications are addition or rewiring of edges. In the model, the edges are directed, therefore every vertex i is determined by both the in-degree k_i^{in} and out-degree k_i^{out}. The algorithm's rules are as follows

1. With probability p *a new vertex is added to the system*. Edges are drawn according to the preferential attachment rule. The key quantity is the in-degree of the target vertex j. In this case the preferential attachment probability is given by $\Pi(k_j^{in}) = (k_j^{in} + \lambda)$

2. With probability $q = (1-p)$ *a new directed edge is added to the system*. The choice of the end vertices depends upon the out-degree k_i^{out} of the originating vertex i and the in-degree k_j^{in} of the target vertex j. This creation function is assumed to be of the form

$$C(k_i^{out}, k_j^{in}) = (k_j^{in} + \lambda)(k_i^{out} + \mu). \qquad (5.163)$$

Through this general model it is possible to derive different distributions for both in- and out-degree. Furthermore, one can study the correlations between these two quantities.

- It is possible to derive analytically the form of the two distributions $P(k^{in})$ and $P(k^{out})$, they are

$$P(k^{in}) \propto k^{-\gamma^{in}} \rightarrow \gamma^{in} = 2 + p\lambda,$$

$$P(k^{out}) \propto k^{-\gamma^{out}} \rightarrow \gamma^{out} = 1 + \frac{1}{q} + \frac{\mu p}{q}. \qquad (5.164)$$

5.4.3   Ageing processes

The last modification of the Barabási-Albert model that we present here is specifically suited for social networks. Consider, for example, the actor network defined by the set of different actors playing in the same movie. Actors can retire or die, therefore their ability to attract new edges in the network finishes after some time. Similar consideration apply to the networks of scientific citations. After some time papers could become obsolete and are eventually removed from the corpus of future citations. These effects can be put into the model by introducing an ageing effect (Klemm and Eguíluz, 2002). Vertices in the network can be either *active* of *inactive*. In the first state they can still receive edges and modify their state. Otherwise their dynamics is frozen and they no longer take part in the evolution of the system. At any time step, the number m of active vertices is kept constant.

1. growth mechanism remains and new vertices enter the system at any time step. Newcomers are always in the active state.

2. A number m_0 of new edges are drawn between the newcomer vertex and *every one* of the active vertices.

3. One vertex i is selected from the set of active ones. This vertex is deactivated and removed from the evolution of the system.

This happens with a probability

$$P_i^{deact} = \frac{1}{N} \frac{1}{(k_i + a)} = \frac{1}{\sum_{j=1,N_a}(k_j + a)^{-1}} \frac{1}{(k_i + a)}. \qquad (5.165)$$

Where k_i is the degree of vertex i, a is a constant, and $1/N$ is the normalization constant given by $1/N = 1/\sum_{j=1,N_a}(k_j - a)^{-1}$.

We list here some analytical results for this model.

- The degree distribution can computed and it is still a power law $P(k) \propto (k + a)^{-\gamma}$.

 This can be seen by considering the time evolution of the probability of receiving an edge for vertex i. This probability at time step $t + 1$ is the same of time t provided the vertex is not deactivated

$$p^{t+1}(k+1) = [1 - P_i^{deact}]p^t(k) = \left[1 - \frac{1}{N}\frac{1}{(k-a)}\right]p^t(k). \qquad (5.166)$$

If the normalization factor can be considered constant in the steady state (where there is no time evolution and therefore $p^{t+1}(k) = p^t(k) = p(k)$) we have

$$\frac{dp(k)}{dk} = p(k+1) - p(k) = -\left[\frac{1}{N}\frac{1}{(k-a)}\right]p(k). \qquad (5.167)$$

In the above equation we treated the probability as a continuous variable, in this hypothesis we can solve the above differential equation and find

$$p(k) \propto (a + k)^{-1/N}. \qquad (5.168)$$

In the steady state the number of vertices n becomes much larger than the number of active nodes. The overall degree distribution can therefore be approximated by considering only the behaviour of inactive vertices. In this way the $P(k)$ is determined by the change of the probability of receiving edges for an active vertex.

$$P(k) = \frac{dp(k)}{dk} \propto (a + k)^{-1/N-1} \qquad (5.169)$$

where the exponent γ is $\gamma = 1/N + 1$.

- The clustering coefficient of this model is larger than that of random graphs and fits nicely the data of some real networks. An analytical estimate gives the value $C = 5/6$ while from computer simulations we find $C = 0.83$.

5.5   Copying models

An example of how a specific case study could inspire the definition of a network model is again given by the World Wide Web. If you want to add your web page to the system (i.e. add a vertex and some edges to the graph) one common procedure is to take one template (a page that you like) and to modify it a little bit. In this way most of the old hyperlinks are kept (Kumar *et al.*, 1999*a*). Maybe new contents are added and eventually some old hyperlinks are kept and some new ones acquired (Aiello, Chung, and Lu, 2000). In the completely different context of protein interaction networks (Vázquez *et al.*, 2003), the same mechanism is in agreement with the current view of genome evolution. When organisms reproduce, the duplication of their DNA is accompanied by mutations. Those mutations can sometimes entail a complete duplication of a gene. A protein can now be produced by two different copies of the same gene; this means that point-like mutations on one of them can accumulate at a rate faster than normal, since a weaker selection pressure is applied. Consequently, proteins with new properties can arise by this process. The new proteins arising by this mechanism share many *physico-chemical* properties with their ancestors. As a result, many interactions remain unchanged, some are lost, and some are acquired.

This growth process works by replicating (with some tolerance) nodes and relative edges already present in the graph. The same mechanism can then be used to define a *copying model*. The rule is local and no global knowledge of the network is needed. This means that we do not have to know the degree of all the vertices to decide how to grow the graph. The above mechanism can be stylized in a model in the following way. At every time-step a randomly chosen vertex is duplicated at random. Each of its m_0 out-going connections is either kept with probability $1 - \alpha$ or it is rewired with probability α.

☕ The rate of change of the in-degree of a node is then given by

$$\frac{\partial k_{in,i}(t)}{\partial t} = (1 - \alpha)\frac{k_{in,i}(t)}{n} + m_0\frac{\alpha}{n} \qquad (5.170)$$

where the first term on the right-hand side of eqn 5.170 is the probability that a vertex pointing to vertex i is duplicated and its edges toward i retained. The second term on the right-hand side represents the probability that the duplicated vertex points toward i by one of its rewired out-going edges. For linearly growing networks we have that $n \simeq t$. The solution of eqn 5.170 is

$$k_i^{in}(t) = \frac{m_0\alpha}{1 - \alpha}\left[\left(\frac{t}{t_i}\right)^{1-\alpha} - 1\right] \qquad (5.171)$$

where t_i is the time when vertex i has entered the network and $k_i^{in}(t_i) = 0$ is the initial condition used to solve eqn 5.170. From eqn 5.171 it is possible to show finally that $P(k^{in}) \sim [k^{in} + m\alpha/(1 - \alpha)]^{-(2-\alpha)/(1-\alpha)}$.

Extensive numerical simulations of such models have confirmed that the resulting **degree distribution** is scale-free. The onset of such scale-invariance is related to the preferential attachment mechanism of the Barabási-Albert model. This can be seen by considering that at every timestep a vertex is chosen at random and its connections are duplicated. Since any vertex can be chosen, the probability of being a neighbour of a vertex of degree k is k/N, where N is the number of nodes in the network. Therefore the probability that a vertex increases its degree (by a unit in a timestep) is proportional to the degree itself. The preferential attachment rule emerges at an effective level from local principles.

In the case of non-directed networks the first term on the right hand side in eqn 5.170 does not change, whereas the second term depends on the degree of the duplicated vertex. The resulting networks are always characterized by heavy-tail degree distributions, although not necessarily strictly scale-free.

5.6   Fitness based model

Although in some contexts preferential attachment can be a very reasonable assumption, in many others it is certainly not. Instead, it is reasonable to think that two vertices become connected when the edge creates a mutual benefit. This benefit depends on some intrinsic properties (authoritativeness, friendship, social success, scientific relevance, interaction strength, etc) of the vertices. Therefore, for some of these systems the scale-free behaviour (when existing) could have an origin unrelated to preferential attachment, but rather related to individuality (or fitness) of the vertices[36].

This idea has been formulated in various ways, and given the different names of static model (Goh, Kahng, and Kim, 2001), fitness model (Caldarelli et al., 2002) and hidden variables model (Boguñá and Pastor-Satorras, 2003). Here we follow the more general formulation (Caldarelli et al. 2002; Servedio, Caldarelli, and Buttà 2004) where no particular hypothesis is made about the statistical distribution used. The network-building algorithm is the following:

---

[36] Furthermore in some situations it may also happen that the information about the degree of other vertex might not be available to newcomers.

- Start with n vertices. For every vertex i draw a real number x_i representing the fitness of the vertex. Fitnesses are supposed to measure the importance or rank of the vertex in the graph and they are extracted from a given probability distribution $\rho(x)$.
- For every couple of vertices, i, j, we can draw an edge with a probability given by the *linking function* $f(x_i, x_j)$ depending on the fitnesses of the vertices involved. If the network is not directed the function f is symmetric that is $f(x_i, x_j) = f(x_j, x_i)$.

A trivial realization of the above rules is the model of Erdős and Rényi (1961). In this case the $f(x_i, x_j)$ is constant and equal to p for all vertex couples. While this particular choice does not produce scale-free networks, as soon as random fitnesses are introduced, the situation changes completely. This model can be considered static as well as dynamic. If the size of the graph is fixed, one checks all the possible couples of vertices as in the random graph model. Otherwise by adding new vertices at every time step, one can connect the new ones to the old ones. A general expression for the **degree distribution** $P(k)$ can be derived easily. Indeed, the mean degree of a vertex of fitness x is simply

$$k(x) = n \int_0^\infty f(x, y)\rho(y)dy = nF(x) \tag{5.172}$$

where x_i belongs to the interval between 0 and infinity. Assuming $F(x)$ to be a monotonous function of x, and for large enough n, we have the relation

$$P(k) = \rho\left[F^{-1}\left(\frac{k}{n}\right)\right]\frac{d}{dk}F^{-1}\left(\frac{k}{n}\right). \tag{5.173}$$

For finite values of n corrections to this equation emerge (Krapivski, Rodgers, and Redner, 2001).

Similarly it can be shown that the average neighbours degree has the form

$$\langle k_{nn}\rangle(x) = \frac{n}{k(x)}\int_0^\infty f(x, y)k(y)\rho(y)dy \tag{5.174}$$

and the clustering coefficient is given by

$$C(x) = \left(\frac{n}{k(x)}\right)^2\int_0^\infty\int_0^\infty f(x, y)f(y, z)f(z, x)\rho(y)\rho(z)dydz \tag{5.175}$$

As a particular example, consider $f(x_i, x_j) = (x_i x_j)/x_M^2$ where x_M is the largest value of x in the network. Then

$$k(x) = \frac{nx}{x_M^2} \int_0^\infty y\rho(y)dy = n\frac{\langle x \rangle x}{x_M^2} \qquad (5.176)$$

and we have the relation

$$P(k) = \frac{x_M^2}{n\langle x \rangle} \rho\left(\frac{x_M^2}{n\langle x \rangle} k\right). \qquad (5.177)$$

The computation becomes simpler if we consider power-law distributed fitnesses. This choice is justified by the widespread presence of scale-invariance in nature (see Chapter 4). If the fitness has the meaning of a ranking, the choice of a power-law distribution for the fitness corresponds to using a Zipf's law.[37] The reason for the ubiquitous presence of Zipf's law relies on the multiplicative nature of the intrinsic fluctuations (see Section 3.6).

Clearly, if $\rho(x) \sim x^{-\beta}$ (Zipf's behaviour, with $\beta = 1 + 1/\alpha$) then, using eqn 5.177, the degree distribution $P(k)$ is also a power-law and the network shows scale-free behaviour. Note that this not a trivial result, the model predicts that if a network is scale-free it can inherit its scale-invariance from the fractal properties of the vertices. This is an interesting conclusion that sheds some light on the tight connection between scale-free networks and scale-invariant phenomena. Whenever Zipf's law is involved, this is enough to create a scale-free network. If this were not enough we mention that one can also obtain scale-free networks when $\rho(x)$ is not scale-invariant. We can consider, for example, an exponential distribution of fitnesses, $\rho(x) = e^{-x}$ (representing a random, exponential distribution) and $f(x_i, x_j) = \theta(x_i + x_j - z)$, where $\theta(x)$ is the usual Heaviside step function that is 1 when x is greater than 0 and 0 otherwise. This case represents a growth of the graph obtained by connecting two vertices if the sum of their fitnesses is larger than a given threshold z. Using these rules, it is possible to check that $P(k) \sim k^{-2}$, as shown in Fig. 5.5. This leads to the non-trivial result that even non scale-free fitness distributions can generate scale-free networks. Different implementations of the threshold rule, such as $f(x_i, x_j) = \theta(x_i^n + x_j^n - z^n)$ (where n is an integer number) give rise to the same inverse square behaviour (although, in some cases, with logarithmic corrections).

As regards the other quantities, we have that

---

[37] We remember here that this law states that the rank $R(n)$ of individual in a population behaves as $R(n) \propto n^{-\alpha}$ in a quite universal fashion (Zipf, 1949).

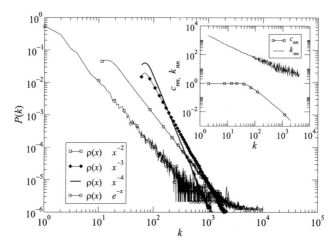

Fig. 5.5 The degree distribution obtained for networks whose fitness is power-law and exponentially distributed. In the inset there is the clustering coefficient and the average degree of neighbours with respect to the degree, in the case of exponential fitness distribution. The disassortative nature of the graph is clear.

- The **distance distribution** $P(d)$ turns out to have small-world properties.
- The **average neighbour connectivity** $k_{nn}(k)$, measuring the average degree of vertices neighbour of a k-degree vertex and the **clustering coefficient** $c(k)$ that measures the degree of interconnectivity of nearest neighbours of k-degree vertices show a disassortative behaviour. In particular, a power-law behaviour is found for $c(k)$ and $\langle k_{nn} \rangle$, when $\rho(x) = e^{-x}$ with a threshold rule.

  In the case of the linking function given by $f(x, y) \propto xy$ it can be shown that these two quantities do not depend upon k as in the case of the Barabási-Albert Model.
- The probability distribution of the **betweenness**, b_i, decays as a power law with an exponent $\gamma_b \approx 2.2$ for $\gamma = 2.5$ and $\gamma = 3$, and $\gamma_b \approx 2.6$ for $\gamma = 4$. This is in good agreement with what was conjectured by Goh, Kahng, and Kim (2001). The exponential case behaves in a different way: for a network of size $N = 10^4$, $z = 10$, and $m = N$ the average distance is $\langle d \rangle = 2$, the total clustering coefficient is $\langle C \rangle \simeq 0.1$ and $\langle b \rangle / N \simeq 0.1$, The betweenness distribution instead shows an unexpected behaviour, giving a power-law tail with an exponent $\gamma_b \approx 1.45$.

Another way to form a network statically, is to start from a given set of connections. Once the number n of the vertices and a degree sequence $k_1, k_2, ...k_n$ are assigned, we can form a graph. The edges are drawn by joining pairs of nodes until every vertex i has degree k_i. Of course there are very strict conditions to be considered in the construction procedure. Firstly, the sum of the degree extracted must be an even number, since it must give twice the number of edges. Secondly, if self and double edges are forbidden,[38] not all random sequences produce acceptable networks. Therefore the procedure must be repeated until all the constraints are fulfilled. We note that for scale-free networks with exponents $\gamma < 3$, it is almost impossible to produce a network avoiding self and double edges.

The procedure outlined above, although very effective, also has the drawback that it does not allow us to change the size of the network during its construction. Any new vertex at the end finds no vertices available to share edges. Therefore the final state of the network is locked.

5.7   Graph from optimization principles

Another way of forming a graph is to shape it according to some cost function. This concept has already been described in Section 4.2 for the evolution of transportation trees. The idea is to fix the number of vertices and edges and to assign a numerical value (hereafter an *energy*) for all the graphs that can be realized with these constraints. If this value has some physical sense (e.g. the average value of distance between vertices) one can select the graphs with the minimal energy and see if they are characterized by some special topology.

For every case of interest this procedure involves an extremely large number of possible graphs and therefore the search for the minimum must be done numerically. This class of search is called the Metropolis algorithm (Metropolis *et al.*, 1953) and works in this way. A starting configuration A is chosen and its energy E_A is computed. Then we make a little change in the graph, this results in a new graph B with an energy E_B. If the energy E_B is smaller than E_A, we accept the change. If the energy E_B is larger than E_A, we could accept the change with a probability $e^{-(E_B-E_A)}$ that is related to the difference $E_B - E_A$. Otherwise, if the change is accepted, the graph B becomes the starting configuration and new modifications are attempted.

---

[38] That is to say we are not considering multigraphs. In a multigraph you have more than one edge between two vertices and a vertex can be connected to itself.

After a while, this recursive procedure selects the configurations with the minimal value of the energy and ultimately the best one.

In this case we can start from a random graph and check what happens by making successive changes in the edge structure. Those changes are driven by our specific request. For example, one could think of reducing (Ferrer i Cancho and Solé, 2003) the vertex-vertex distance. To this aim we define a quantity measuring the density of the network.

$$\rho = \frac{\langle k \rangle}{n - 1} \qquad (5.178)$$

where $\langle k \rangle$ is the average value of the degree and n is the order (number of vertices) of the network. From the definition it turns out that the maximum theoretical value of ρ is one, and typical values for ρ in real networks range from $\rho_{min} = 10^{-5}$ to $\rho_{max} = 0.5$ (World Trade Web).

At this point we can measure the optimization of a network by assigning to any graph a function

$$E(\lambda) = \lambda d + (1 - \lambda)\rho \qquad (5.179)$$

where d is the normalized distance between two vertices. This latter quantity is given by

$$d = \frac{\langle D \rangle}{D_{linear}} = \frac{\langle D \rangle}{(n + 1)/3} \qquad (5.180)$$

where on the numerator we have the average value of the distance between all the possible couples of vertices and on the denominator we have the maximum of this value that is obtained for the so-called 'linear graph'. A linear graph is a graph having the maximum distance possible. On this lines it has been analytically demonstrated that when the total length is fixed, the optimal network which minimizes both the total length and the diameter lies in between the scale-free and spatial networks (Barthélemy, 2003).

This is one simple way of imposing a cost function on the structure. Others are possible, as shown in the next section.

5.7.1   Optimized graphs: cost function and transport

Sometimes (and this is the most interesting case) the graph is a structure designed in order to realize in an optimal way some specific functions. This class of graphs are called 'economic' graphs in the language of graph theory. Trees are usually very good candidates as economic graphs, because in a tree we connect n vertices with the minimum possible number of edges $(n - 1)$.

The concept of an economic graph can be related to 'dynamics' as a transport along the edges of the graph. One immediate example of a transportation network could be represented by a pipeline connecting different houses. Houses are the vertices of the graph, while the pipes are the edges. Other examples could be cities connected by railways or houses served by electricity cables. We focus on another important and interesting case within our bodies, namely the network composed by our blood vessels (see Fig. 5.6).

We show now that in this case this transportation network is optimized. This means it has the best possible shape to accomplish its function. More specifically, it delivers the nutrients to the body with the minimum possible amount of matter transported. Not only blood vessels but also river networks and plant channels deliver their contents in the most efficient way. We report here some of the work done on transport optimization. The results hold for a lattice of any Euclidean dimension d and we present here an extension of this result for graphs not embedded in Euclidean space. The physical motivation for that research was aimed at explaining the Kleiber relation, which states that the metabolic rate B of every animal grows with its body mass M with a specific exponent[39]

$$B \propto M^{3/4}. \tag{5.181}$$

This relation holds for different species (West, Brown, and Enquist, 1999; Banavar, Maritan, and Rinaldo, 1999). Note that the scaling is not *isometric*: (i.e. for compact objects the metabolic rate B should increase as $M^{2/3}$). Rather it is an *allometric* relation (i.e. B and M grow with a different relation).

As far as we are concerned, we consider here the various species characterized by a volume (the body) served by a network (blood vessels). The above result can be explained by showing that the network of blood distribution is a particular kind of tree. To show that, we proceed as follows

- We select one source in the network (which can be regarded as the root of the transportation tree) from which the flow proceeds.
- We compute the distance of any other vertex from the root. This procedure is also known as breadth-first search (see Section 1.2.1).

At this point we have created a tree out of the network. This tree can be shaped in different ways, some limit cases are shown in Fig. 5.7. For the moment let us not consider that and let us focus only on the topological meanings of the eqn 5.181.

---

[39] For our purposes we can think of the metabolic rate as the number of calories an organism needs to survive.

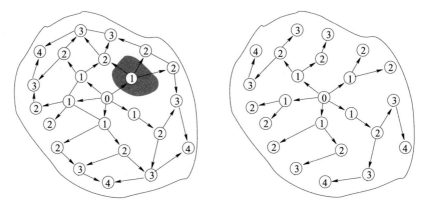

Fig. 5.6 Left: a fully connected network (for example a portion of a body with its blood vessels). The grey portion indicates the region served by the central vertex. Right: the tree obtained by measuring the distance from supplying vertex. This distance is indicated in the centre of the vertex.

An animal body is described as a system of length L embedded in a space with Euclidean dimension $d = 3$ (our familiar three-dimensional space).[40] If we represent the organism as a d-dimensional lattice of L^d vertices with unit spacing, we have that every vertex v out of the total L^d is supplied with a flow F_v at a constant rate. The metabolic rate of the organism is given by

$$B = \sum_v F_v \qquad (5.182)$$

where F_v is the metabolic rate of vertex v. *The metabolic rate B grows as L^d when the size L grows.* This is because B is the product of a nearly constant term F_v times the number L^d of the vertices.

Similarly we can compute the total volume of blood C used in the organism (this is the counterpart of M in eqn 5.181. If we denote by I_e the flow of nutrient on every edge we have that this quantity C is given by

$$C = \sum_e I_e \qquad (5.183)$$

where it is summed over every edge e. The function C plays a crucial role in the study of the *allometric relations*. Different results are possible according to the kind of the tree (Banavar, Maritan, and Rinaldo, 1999). In the first

---

[40] This latter value can change, the dimension of the embedding space is 2 for a river network since water flows on a surface. It is almost 2 for a thin leaf, which is the space within which the nutrients flow.

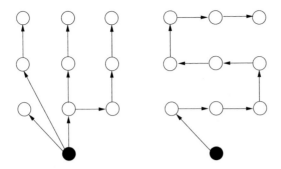

Fig. 5.7 Left: an optimized tree whose C grows as L^{d+1}. Right: a tree whose C grows as L^{2d}.

case shown in Fig. 5.7 *the quantity C grows with L as L^{d+1}*. The same quantity grows as L^{2d} in the second case. Since C represents the mass, the first choice is the optimal one in the sense that it reduces the amount of mass necessary to deliver nutrients to the same area. Not surprisingly, nature selects the optimal solution and in the first case we recover the Kleiber relation.

$$
\left.
\begin{aligned}
B &\propto L^d \\
C &= M \propto L^{d+1}
\end{aligned}
\right\} \rightarrow B = M^{\frac{d}{d+1}}.
\tag{5.184}
$$

Which in the case of $d = 3$ gives exactly the exponent 3/4. We do not demonstrate here that $C \propto L^{d+1}$ is the best choice. We only note that in order to feed every one of the L^d vertices the blood has to travel at least the mean distance (of order L).[41] This gives exactly the exponent $d+1$ claimed to be the optimal one.

5.7.2   The constant flux

Let us consider a very specific case. The flow I_e is constant to every 'leaf' (i.e the vertices with only one edge). Without loss of generality we can consider this I_e unitary. It is easy to check by recursion that the role of the metabolic rate is played by the drained area a (or the basin size n) introduced in Section 1.2.

$$
A = \sum_{i \in \; basin} a_i
\tag{5.185}
$$

where we consider all the sub-basins of area a_i and outlet i present in a certain basin (see Fig. 1.8). This is because summing on all the unitary fluxes

---

[41] Note that a graph not embedded in a Euclidean space has small-world properties, so that one can expect the mean distance to grow with the logarithm of L.

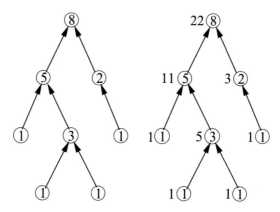

Fig. 5.8 Values of A (left) and $C(A)$ (right) on a simple tree. Consider the blue vertex in this tree. On the right for the computation of $C(A)$ we sum all the values in the subtree, as reported on the left. This results in a value $5 = 3 + 1 + 1$.

corresponds to summing on all the points in the sub-basin. One possible global measure for the shape of such a quantity is given by the probability distribution $P(A)$ of finding a basin of a certain size A in the system.

Consider now the case of a subtree of area A. For every subtree we can compute the quantity $C(A)$. In the case of constant and unitary flux, the total volume of blood (or water or whatever) is simply given by the sum of the values A in the (sub-)basin.

In formulas we can therefore write

$$C(A) = \sum_{i \in A} A_i \qquad (5.186)$$

where the sum is over on all the vertices i (of area A_i) in the sub-basin of area A.

The real meaning of these quantities is related to the particular cases where they appear (e.g. in river networks and biology). As shown in the next chapters, real river networks are all characterized by the same shape of $P(A)$. Similarly, biological and other natural systems are characterized by the same $C(A)$.

PART II

EXAMPLES

6.   Networks in the cell

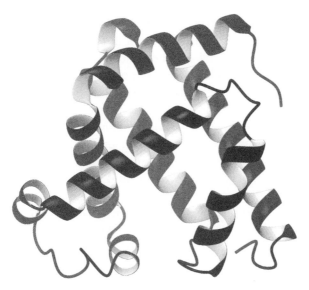

Fig. 6.1 The three-dimensional structure of myoglobin, which allows the transfer of oxygen in the body.

If you really want to hear about it, the first thing you'll probably want to know is if the concepts of scale-free networks can be useful for cell biology. Actually, there are various biological processes within a cell where we can apply the concepts presented in the previous chapters. Unfortunately in most of the biological cases (at least presently) large sets of data comparable to those of the Internet and the WWW are not available. However, due the relevance of this field, many new experiments are being conducted worldwide. For this reason the amount of experimental evidence is starting to be rather impressive and the data more and more reliable.

The basic players of this chapter are proteins (see Fig. 6.1) and nucleic acids. Proteins are one of the basic constituents of life. In our body they are usually folded in some particular shape. They are polymers (long chains of smaller blocks), formed by a sequence of aminoacids taken from a repertoire of no more than twenty different basic building blocks. The information about the composition of proteins needed by an organism is stored in another macromolecule: a nucleic acid called DNA. This macromolecule is also a sequence made of no more than four building blocks named *bases* (Watson and Crick, 1953).

Even though recent research has started to challenge the idea, we make explicit use in this chapter of the so-called 'sequence hypothesis' and the 'central dogma of cell biology' (Crick, 1958). We then assume that the specificity of a piece of nucleic acid lies solely in the sequence of its bases and that sequence is a code for the amino-acid sequence of a particular protein. The way in which aminoacids are collected in the right order and the protein is formed is made possible by another nucleic acid called RNA (also composed of only four bases). The central dogma (shown in Fig. 6.2) states that the precise determination of a protein sequence is coded only in the DNA and cannot be changed or influenced by other proteins. If the code of this sequence is called information this means that 'the transfer of information from nucleic acid to nucleic acid, or from nucleic acid to protein may be possible, but transfer from protein to protein, or from protein to nucleic acid is impossible' (Crick, 1958; Watson and Crick, 1953). This statement supported by more experimental evidence was again considered after some time (Crick, 1970). Despite the existence in nature of some special exceptions, we restrict ourselves only to cases where the central dogma holds. It is important to note that this does not means that proteins cannot act on nucleic acid. Proteins can interact with nucleic acids, making possible a series of biochemical reactions that ultimately might trigger the production of new proteins. The dogma simply states that proteins cannot form new parts of nucleic acid. On the contrary nucleic acids can form new proteins according to their recipes. Therefore the only source of information is that coded in the DNA.

The above series of reactions is the first case of a network in the cell, that of the protein–protein interaction network. In this network the vertices are the proteins and an edge is drawn if two of them physically interact in the same biological pathway. Since most of these functions have not been completely understood, nor the function of every protein known, a network approach is likely to help in the understanding of the phenomenon. If two proteins interact with one another, they usually participate in the same, or related, cellular functions. Protein interaction networks therefore can be a way of determining how proteins physically interact within a cell. To observe or not this interaction is not trivial at all. In most of the cases the investigation is done through a method known as 'two hybrid' (Chien

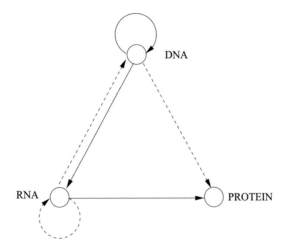

Fig. 6.2 The flow of information in gene expression after Crick (1970). Solid lines are flow of information for which evidence was present at the time the paper was written. Dashed lines correspond to hypothetical information. Since the original formulation of the central dogma (Crick, 1958), the status of the RNA loop changed from evident to hypothetical. No flow of information is evident from proteins to DNA or RNA.

et al., 1991). This is an indirect method of testing protein interactions which is applied to thousands of protein pairs at the same time. Despite many efforts, this method is still far from providing reliable sets of data even for very simple organisms like the baker's yeast *Saccharomyces cerevisiae*. Different groups working on this subject have produced different maps of interaction whose results coincide only for a small percentage of the number of interactions.

Another application of graph theory within the cell is that of metabolic networks. Metabolism refers to all the chemical changes taking place in an organism in order to produce energy and basic materials needed for important life processes. In this case the vertices are substrates, that is the reactants and products of chemical reactions. Those chemical reactions are the edges of the graph. It can be 'anabolism' when we have synthesis of larger molecules from simpler ones (e.g. amino-acid synthesis). We have 'catabolism' when we break down complex molecules to simpler, smaller molecules (e.g. glycolysis). The molecules involved in this process form a network made of vertices (metabolites) connected through edges (biochemical reactions). In general we say that we have a change from substrates S into products P ($S \rightarrow P$). This can be done spontaneously or catalysed by specific enzymes E ($S + E \rightarrow P$). An enzyme is biological molecule (often another protein) acting as catalyst in a chemical reaction, thereby enhancing the rate of the reaction without being consumed in the process. This process is very general with most of the biochemical reactions represented by two-way processes, catalysed by enzymes. This clarifies that nucleic acids play a role even if indirectly also in this case. The protein synthesis of structural genes produces specific enzymes catalysing biochemical reactions. The transcription of these genes is also to be regulated by enzymatic mechanisms.

This introduces the third case study, that of gene regulatory networks. A cell must change its composition continuously in order to work properly. Furthermore (as any living organism) it has to face different external conditions and react to them to survive. Not all the genes are expressed at the same time; rather there is a precise schedule of activation and expression. In this mechanism the genes (subparts of DNA) can be viewed as nodes in a network, and an edge is drawn between two vertices whose actions are causally connected.

> In this chapter we present three of the most studied networks in cell biology. To present them and to describe the nature of these network we also briefly introduce the basis of cell biology. Activity in this field is frantic; therefore we can guess that in a few years, more data will be available and the general understanding will be clearer. In general all these networks present some of the properties of scale-free networks. Application of modelling and concepts of scale-free networks can trigger further understanding.

6.1   Basic cell biology

Cell biology is so complicated, that even a basic summary could occupy a book much larger than this. Consider this paragraph only as a stimulus to your interest and to drive your attention to more specific books (e.g. Nelson and Cox 2005). Many questions are open and many things are left to be completely understood. Here we restrict as much as possible to three basic structures: the proteins, DNA, and RNA. These three ingredients play a crucial role both in prokaryotic species (single-celled individuals) and eukaryotic species (the more recent multicellular individuals).

6.1.1   The recipe of proteins: the DNA

The instructions for building a protein are coded in the genetic code of organisms by means of another molecule, DNA. DNA is the acronym for deoxyribonucleic acid. Both the coding and the replication are accomplished by the characteristic structure of this molecule. The physical form of DNA is given by two different strands. These strands are composed of only four different building blocks called nucleotides. Those nucleotides, adenine (A), thymine (T), cytosine (C) and guanine (G) have a particular and crucial property. 'A' binds only with 'T' and vice versa, 'G' binds only with 'C' and vice versa. Therefore given one strand we know what is present on the other. This is the way in which this molecule can replicate itself. Once the two strands of the DNA are separated, they can collect from the external environment the precise nucleotides, forming again the missing strand. In this process we obtain two copies of the original molecule (actually it is a little bit more complicated than that). These strands are held together by hydrogen bonds[42] between nucleotides G and C and nucleotides A and T. These two strands entwine like vines to form a double helix,[43] as shown in Fig. 6.3. While most of the functionality of this sequence is still unknown, it is now clear that parts of the sequence trigger the formation of proteins through RNA. The relationship between the nucleotide sequence and the amino-acid sequence of the protein is determined by simple cellular rules of translation, known collectively as the genetic code.

6.1.2   The building of a protein: RNA

Proteins are not formed by directly assembling molecules on the DNA strand. Rather a little part of the gene is transcribed to another molecule called RNA. Through a series of complex chemical reactions, happening in a different part of the cell, RNA collects the series of amino acids necessary to form the protein.

The process of copying DNA to RNA is made by an enzyme called RNA polymerase (RNAP). The process is rather complicated and differs between prokaryotes and eukaryotes. In the latter case different enzymes for RNA polymerase form several types of RNA molecules. In principle, the mechanism is the same as the simplest one in prokaryotes where a single species of

---

[42] That is crucial because hydrogen bonds are strong enough to keep the molecule together, but weak enough to allow local splitting of the strands.

[43] In prokaryotes and bacteria the DNA is within the cell of the organism. In eukaryotes (organisms with nucleated cells, as animals plants and fungii) the DNA is confined within the cell nucleus and is composed of different continuous chains of nucleotides called chromosomes. In this latter case the 'DNA' is the series of different chromosomes.

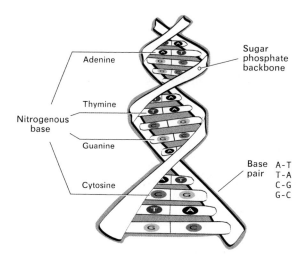

Fig. 6.3 The structure of DNA crucial for its replication process.

enzymes mediates the RNA synthesis. In general the polymerase starts the reaction separating the strands on starting points. While the production of RNA proceeds, the DNA binding between the strands reforms behind the RNA. This process releases the RNA chain from contact with DNA.

The chemical nature of RNA is different from DNA, having a different sugar molecule (ribose instead of deoxyribose), the bases forming RNA are similar to DNA apart from thymine (T), which is replaced by uracil (U). In prokaryotes the RNA formed in such a way is able to start the protein production. In this case the RNA is called messenger RNA (mRNA). In eukaryotes the process happens in the ribosome. This is a part of the cell made of RNA and ribosomal proteins. Here the messenger RNA (mRNA) is translated into a polypeptide chain (e.g. a protein).

6.1.3   Protein structure

Cells are made of many constituents, water, minerals, lipids to form the external membranes, and elaborate organic compounds including proteins. They account for more than 50 per cent of the dry weight of most cells. Most of the life functions are carried on by proteins and even their name (it comes from the ancient Greek) means 'basic' since they were believed to be the elementary structure in the process of organism nutrition.[44] As DNA is made of only four nucleotides, the proteins are sequences (up to more than

---

[44] This name was proposed in 1838 by scientist Jöns Jacob Berzelius.

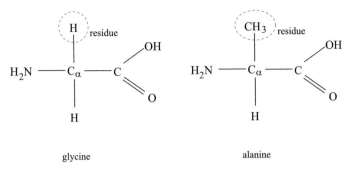

Fig. 6.4 Left: the simplest possible amino acid, *glycine*. Glycine residue is composed of a hydrogen atom H. Right: a slightly more complicated amino acid, *alanine*, whose side residue is made of a methyl group CH_3.

25,000 units) of only twenty *amino acids*. Amino acids are a particular class of molecules sharing a similar structure. They consist of a central carbon atom (the carbon C_α) an amino group (NH_2), a hydrogen atom (H), a carboxyl group (COOH) and a side chain (R) which are bound to the C_α. Different side chains (R_i) form different amino acids with different physical and chemical properties. The 'alpha-carbon' is a name given by convention to the central carbon atom. Both the amino and carboxyl groups are covalently linked to the alpha-carbon.

The simplest and smallest of the standard amino acids is *glycine*, where the residue is composed only of a hydrogen atom (R_{gly}=H). *Alanine* is slightly more complex in terms of its structure and chemistry as compared to glycine; both are shown in Fig. 6.4. Now the residue is composed of a CH_3 molecule (methyl group, in this case R_{ala}=CH_3) . Obviously, different amino acids exhibit different chemical and/or physical properties according to their corresponding R-groups. In particular conditions different amino acids can be grouped forming a larger molecule (the first step in protein formation). The basic mechanism is the formation of a peptide bond with the loss of a water molecule, as shown in Fig. 6.5. A peptide bond is formed by binding covalently a carbon atom of the carboxyl group of one amino acid to the nitrogen atom (of the amino group) of another amino acid. This process happens by dehydration, that is, one molecule of water is produced in the reaction. The newly created C-N 'peptide' bond bridges the two separate amino acids. A chain of molecules connected by peptide bonds is also called a polypeptide. A protein is made up of one or more polypeptide chains (called domains), all made of amino acids. Once they are produced they fold in particular geometrical shapes and are ready to perform a particular task.

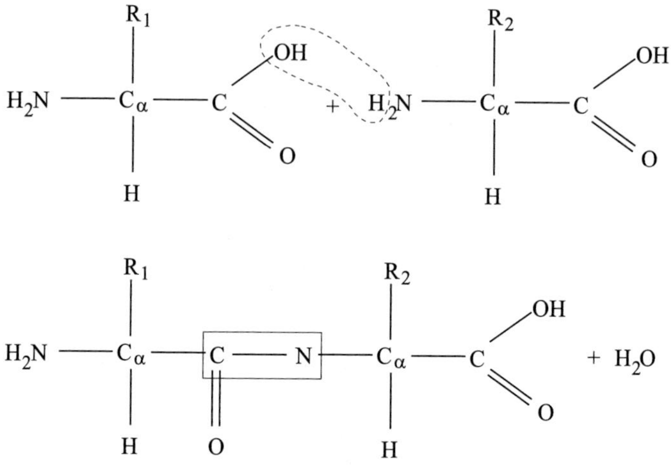

Fig. 6.5 The basic reaction forming a peptide bond by dehydration.

The shape into which a protein naturally folds is unique (native state) and is determined by the sequence of amino acids.

Some proteins can act as enzymes, that is to say they make possible a chemical reaction (*catalysis process*). This feature will be particularly important in the case of metabolic pathways as shown in the following.

6.2   Protein–protein interaction network

The functionality of a cell is a very complicated process far from being completely understood. Nevertheless, in this system a key role is played by interactions of different proteins binding to each other. This system allows for a simple graph representation where the vertices are the different proteins and an edge is a physical interaction between the two. This would mean, for example, considering two proteins as interacting if they participate in the same complex (i.e. a macromolecule composed by several proteins).

Apparently from different analysis these networks show the typical structure of scale-free networks with 'hubs' that is some proteins interacting with many partners and many interacting with only few of them. This structure seems to be connected with the importance of the protein for development of cells. It has been shown that deletion of these hubs (at least for *S. cerevisiae*) is lethal for the organism. This means that proteins with large degree are more likely to be more essential than proteins with low degree (Jeong

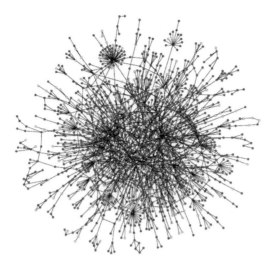

Fig. 6.6 The graph of the protein interaction network of *Saccharomyces cere-visiae*. Courtesy of H. Jeong (Jeong *et al.*, 2001).

et al., 2001). Note that proteins with large degrees can interact simultane-ously with other partners or interact at different times or locations (Han *et al.*, 2004).

At the moment a rather large amount of information and data is available. As time passes, new organisms are considered and new data sets are pre-sented to scientific community. Furthermore, careful new analyses help in the determination of the validity of older measurements. At the time of writing, many data sets are present for baker's yeast *Saccharomyces cerevisiae*[45] and for *Drosophyla melanogaster* (Giot *et al.*, 2003). The first analyses (Jeong *et al.* (2001); Wagner (2001)) use data collected on this organism, using a two-hybrid method (Uetz *et al.*, 2000). In the second of the above papers we have a network composed of 899 interactions for 985 proteins. Since the number of edges is less than those of vertices, this means there are some disconnected parts. Actually the network is composed of as many as 163 components. The giant one is formed by 466 vertices and 536 edges. In this one the degree is power-law distributed with an exponent of about 2.5. Other studies (Ito *et al.*, 2001; Maslov and Sneppen, 2002) report 4,549 physical interactions (the edges) between 3,278 proteins (the vertices). The degree distribution is again power law with a similar exponent $\gamma = 2.5 \pm 0.3$. From

---

[45] http://www.yeastgenome.org/VL-yeast.html

	Size	Order	γ	$\langle k \rangle$	Source
S. Cerevisiae 1	899	985	2.5	1.83	(Wagner, 2001)
S. Cerevisiae 2	4549	3287	2.5	2.63	(Ito *et al.*, 2001)
E. Coli	716	233	1.3	6.15	(Butland *et al.*, 2005)
D. Melanogaster	4780	4679	1.6	2.05	(Giot *et al.*, 2003)

Table 6.1 Basic table for protein interaction networks. When not indicated in the original papers the value of $\langle k \rangle$ is computed from the first two columns. In the case of *Escherichia coli*, the authors do not report the value of the exponent γ. From a rough estimate of two decades on y and one and half for axis x we obtain the value presented here. In case of *Drosophila* the authors fit all the data with a power law plus an exponential cut-off. This is in general always the real case and whenever discussing power laws we refer in this book only to the linear part. That linear part as can be inferred from the picture and gives a value of about $\gamma = 1.6$.

a plot of correlation it is also present a disassortative behaviour for the system. This means that there is a tendency for poorly connected proteins to be connected to large hubs rather than each other. This is also confirmed by a study of the average connectivity $\langle k_{nn} \rangle$ of the neighbours of a vertex whose degree is k. This is a decreasing function of the kind $\langle k_{nn} \rangle (k) \propto k^{-0.5}$.

In the case of the fruit-fly protein–protein interaction network (Giot *et al.*, 2003), the interactions detected through the two-hybrids method are also weighted with the frequency of observation. By putting a threshold on this confidence value the authors were able to obtain two distinct networks: a general one and a 'high confidence' one. The set of proteins detected are in the first case 7,048 connected by a series of 20,405 interactions (edges). In the case of the high confidence network the size of the graph reduces to 4,780 edges and the order to only 4,679 vertices. As regards the component of the network, it consists of a giant connected cluster (3,039 proteins, 3,659 interactions) and 565 smaller clusters (about 2.8 proteins and about 2.0 interactions per cluster on average). The results show a small-world effect even if the probability distribution for distances is peaked at a large value of 9.11 edges; the mean is nearby with a value of 9.4 edges. In an ensemble of randomly rewired networks of the same size, the mean separation is about 7.7 edges. Clustering is computed considering the nature of cycles (triangles, squares, pentagons, etc.) in which the perimeter is formed by a series of proteins connected head-to-tail, with no protein repeated. The actual network shows an enhancement of loops with perimeter up to 10.11 relative to the random networks. Giot *et al.* (2003) present a simple model to reproduce the properties of the *Drosophila melanogaster* protein–protein interaction network. The whole structure is represented by means of different layers of

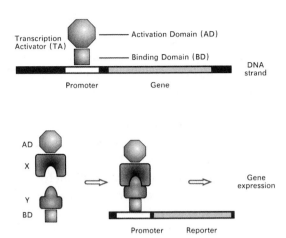

Fig. 6.7 The mechanism of two-hybrid method. AD stands for activating domain and BD stands for binding domain.

organization. At the lower level the model describes protein complexes (i.e. macroproteins made of more than one simple protein), at the upper level we have interconnections of these complexes. A similar conclusion for the hierarchy in complexes has also been found very recently (Butland *et al.*, 2005) analysing the protein–protein interaction network of *Escherichia coli*.

All the above results are summarized in Table 6.1.

6.2.1   The two-hybrid method

This method (Chien *et al.*, 1991) detects interactions between proteins by using the activation of a gene by means of a transcription factor.[46] This transcription factor is split into two separate pieces, one part that directly binds into the genetic strand (the *binding domain*), and another that activates the transcription (*activating domain*). Without the two of them acting together, the transcription cannot start and there is no production (see Fig. 6.7). Using this process one then selects two proteins: the *bait* and the *prey*. Bait are usually known proteins for which one would like to know the set of interactions. Prey proteins can be either known proteins or a random library protein (the interest in protein interactions is due to the fact that in this way one can successfully classify new proteins). By genetic engineering

---

[46] The latter is nothing more than a protein that by binding on a specific portion of the genetic code is able to start or increase the transcription of the DNA in the complementary RNA and eventually another protein.

the bait protein is fused with the binding domain. Prey proteins are then fused with the activating domain. At this point if the bait and prey proteins interact by forming a binding, then the two parts of the transcription factor are again bound together and the transcription of gene can start. From the product of such transcriptions one can measure the interaction between prey and bait. It is clear that an indirect method like this can have both false positives (i.e. transcription starts, but only because binding and activating domain are able to interact with each other without binding of prey and bait) and false negatives (i.e. prey and bait do interact, but the binding does not allow us to form a proper transcription factor). For such reasons, detection of interactions is usually confirmed by careful choice of bait and prey couples (Fromont-Racine, Rain, and Legrain 1997; Formstecher *et al.* 2005) or by different methodologies (Engel and Müller, 2000; Dufrêne, 2004; Rossini and Camellini, 1994). In any case in the absence of a reference set of validated protein interactions the quality of these predictions remains uncertain. Another possibility is that even interactions effectively measured have nothing to do with cell functionality. Rather in the spirit of the fitness model (see Section 5.6) we can reproduce most of the above results by taking into account (de)solvation as the key ingredient (Deeds, Ashenberg, and Shakhnovich, 2006). This idea is further supported by the evidence that the protein's degree is positively correlated with the number of hydrophobic residues on the surface.

6.3   Metabolic pathways

Living organisms rarely obtain the specific molecule they need from the environment. Often (if not always) they must transform what is available with a series of complicated chemical reactions. Very roughly, we can say that in certain conditions specific elaborate macromolecules (proteins) must be produced starting from simpler molecules. In other conditions, large external macromolecules can be split obtaining smaller subunits and energy. This energy can be used to sustain other chemical reactions. In this process a particular chemical transformation is made possible or accelerated or maybe stopped by the presence of other chemical compounds called enzymes. We take as *metabolism* the sum of these enzyme-catalysed chemical reactions taking part inside the cell (Nelson and Cox, 2005). For the sake of simplification sometimes a precise series of reactions is referred to as a metabolic pathway. We see in the following that this is a considerable simplification: in the real situation different pathways are interconnected in an intricate network. This situation is very similar to ecology where food chains are

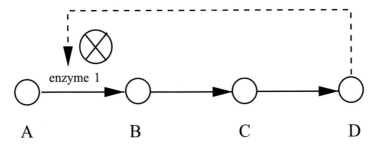

Fig. 6.8 An elementary metabolic pathway. The series of reactions produces different compounds. The last one can act as an enzyme, slowing down production of B and thereby self-regulating its production.

introduced as a first step towards the more real situation of food webs. Some authors explicitly report the various steps in this conceptual passage from simple representation to complex (Papin *et al.*, 2003). Any step in a metabolic pathway is a specific chemical reaction, while the intermediate states from the initial molecule (precursor) to the final one (product) are *metabolites*. One example is glycolysis, which is one of the processes that generates energy in the body. A sugar (glucose) is partially broken down by cells in enzyme reactions that do not need oxygen.

A typical oversimplified series of reaction is that shown in Fig. 6.8. Starting from the precursor A one passes through different metabolites B, C, D. These compounds, rather than being inert. may be enzymes themselves. For example, an elevated concentration of metabolite D can stop production of B. This is a typical case since most of the times these reactions have a feedback mechanism that allows self-regulation of the chemical process. As previously noticed, the series of reaction is a network rather than a chain; this means that the various feedbacks affect other pathways in an almost unpredictable way. Such a structure can be analysed easily using the tools of graph theory, by defining the metabolites as vertices and the reactions as edges. A more complex pathway is that of the carbon central metabolism shown in Fig. 6.9.

Most reactions are not reversible, so that the network is directed.[47] Also these networks present a power-law distributed degree of the kind $P(k^{in}) \propto (k^{in})^{-\gamma^{in}}$ with $\gamma^{in} \simeq 2.22$ (Jeong *et al.*, 2000). Such a structure checked for 43 different organisms implies that some metabolites act as hubs for various different processes and therefore have a particular importance in cell

---

[47] In principle, by knowing the relative concentration of the various compounds it should be possible to obtain a weighted description.

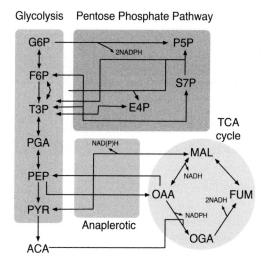

Glycolysis   Pentose Phosphate Pathway

Fig. 6.9 Bioreaction network of *E. coli* central carbon metabolism. The arrows indicate the (assumed) reversibility or irreversibility of the reaction. Fluxes to biomass building blocks are indicated by solid arrows. The abbreviations are: G6P, glucose-6-phosphate; F6P, fructose-6-phosphate; P5P, pentose phosphates; PP pathway, pentose phosphate pathway; E4P, erythrose-4-phosphate; S7P, seduheptulose-7-phosphate; T3P, triose-3-phosphate; PGA, 3-phosphoglycerate; ACA, acetyl-coenzyme A; ACE, acetate; OGA, 2-oxoglutarate; PYR. pyruvate; FUM, fumarate; MAL, malate; OAA, oxaloacetate; and PTS system, PEP:carbohydrate phosphotransferase system for glucose uptake. After Segrè, Vitkup, and Church (2002).

development. Surprisingly, most of the vertices are connected by a relatively short path while some authors found an average path length of about 3.2. This means that most of the metabolites can be transformed to each other in about only three steps. Successive analyses changed this picture only slightly (Ma and Zeng, 2003). In *Escherichia coli* the average path length is still small, but with a larger value of 8.2 and the diameter (i.e. the maximum of distances) is 15. Rather than universality, some difference exists between organisms; eukaryotes and archaea have a longer average path length than bacteria.

A possible explanation for such a discrepancy is in the fact that the first analysis considers ATP and ADP as nodes in the networks. Some others instead separate the process taking place with these molecules giving rise to more paths and generally a larger path length. The various metabolic

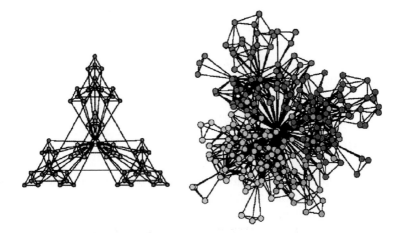

Fig. 6.10 A possible hierarchic structure to explain the large-scale invariance observed in various metabolic networks. Left: we have a sketch of the hierarchical model used. Right: a rearrangement in space with a clustering algorithm. The same colour for vertices indicates the same level in the hierarchy. Courtesy of E. Ravasz (Ravasz *et al.*, 2002).

networks belong to fully sequenced organisms and are reconstructed *in silico*[48] from genome data. One widely used gene catalogue available on-line is the KEGG (http://kegg.genome.ad.jp). For the purposes of classification the various reactions are listed according to the particular pathway they belong to (even if this is a simplification). Other reconstructions of the *S. cerevisiae* metabolic network show a correlation with that of *E. coli* (Förster *et al.*, 2003).

A hierarchical model of network can also explain the statistical properties of metabolic networks (Ravasz *et al.*, 2002). The idea is to realize a fractal assemblage of vertices, as shown in Fig. 6.10. The model takes a basic module of strongly interconnected vertices and generates three replicas of it. The replication is iterated and any time the peripheral nodes are connected with the central one of the starting module. A structure like this is able both to reproduce the scale-free properties of the net as well as the disassortative structure measured from the data. Organisms like *E. coli*, *S. cerevisiae* and *Aquidex aeolicus* all have a clustering coefficient $C(k)$ with a power-law distribution of the kind $C(k) \propto k^{-1}$ as in the model. More importantly,

---

[48] With no hope of producing any noticeable change in the field, we note that the correct form should be '*in silicio*'.

the structure of the model closely resembles the modular structure of the metabolic network of *Escherichia coli* once the various chemical reactions are classified according to traditional classes (i.e. carbohydrate; nucleotide and nucleic acid; protein peptide and amino acid; lipid; aromatic compound; monocarbon compound; coenzyme metabolisms).

6.4   Gene regulatory networks

Not all the proteins a cell can produce are present at the same time, this means that only a fraction of the available genes are expressed. Production of one protein it is a very energy-consuming task and mechanisms of regulation (more or less complicated feedbacks) of gene expression have evolved to produce only the necessary quantity of the required protein at due time. A gene regulatory network can be described as the series of all the chemical reactions involving gene expression in the cell. The intensity and duration of these reactions may also be different within cells in a population and are also affected by external environmental conditions.

This is somewhat different from the previous metabolic network which describes reactions not involving gene expression (and whose main purpose is the production of energy). Expression of one gene is made possible by RNA and other proteins, some authors consider genes, RNA, and proteins as vertices of the networks, others only genes. In the former case one edge is any chemical reaction involving the two vertices, in the second case it is a series of reactions that causally connects expression of one gene with the expression of another.

One of the first descriptions of this system by means of a network (Kauffman, 1969) uses the first approach where vertices can have different nature. In this example, the network is composed of proteins that act as input signalling pathways, regulatory proteins that integrate the input signals, the genes activated, and all the RNA and proteins produced from those target genes. A directed edge connects them if there is a causal connection between them. Vertices can be in two states 'on' (1) or 'off' (0); giving information if the gene is expressed or the protein/RNA are present. The network is dynamically updated and at each time step every vertex assumes a new state that is a Boolean function[49] of the prior states of the vertices pointing towards it.

---

[49] For our purposes, a Boolean function can be defined as a function that combines different variables $(x_1, ..., x_n)$ and returns either 0 or 1.

An important property of the gene regulatory network is feedback. The cell reaction can change the environment, consequently the reaction must be tuned to account for the new situation. The feedback can be positive, that is constant operation of the cell can reinforce a situation, or negative. In the latter case a down-regulation of the process allow a steady state for the cell to be attained (Becskei and Serrano, 2000).

In more complicated situation as the process of meiosis, the cell alternates in two different states. At the beginning it has to double its genome and after that it has to divide into two parts in order to form two different cells. A special feedback is necessary since specific genes must be turned on and off periodically to obtain that result. All these feedbacks form very often a series of cycles. An example of that highly dynamic sequence of activation is present in a study of the bacterium *Caulobacter crescentus* (Laub *et al.*, 2000) The cells of *Caulobacter* use some transcription patterns to control cell cycle progression. Global transcription analysis was used to identify 553 genes (19 per cent of the genome) whose messenger RNA levels varied as a function of the cell cycle. The main results are that genes involved in a given cell function are activated at the time of execution of that function. Furthermore the genes encoding proteins that function in complexes are co-expressed while temporal cascades of gene expression control the formation of multi-proteins. A single regulatory factor, the CtrA (member of the two-component signal transduction family), is directly or indirectly involved in the control of 26 per cent of the cell cycle-regulated genes (suggesting a role as that of a hub of the network). Also in this case it is supposed that a hierarchy of transformations is responsible of the structure of the network. Interestingly, a coarse graining approach clustering several reactions in 'modules' seem also to be a solution for more reliable measurements (Kholodenko *et al.*, 2002).

7. Geophysical networks

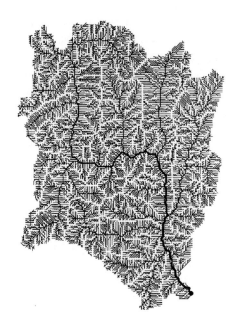

Fig. 7.1 A representation of the river network of the Fella River (tributary of Tagliamento River) in the north-east Italy. Courtesy of I. Rodríguez-Iturbe and A. Rinaldo.

A few miles south of Soledad, the Salinas River drops in close to the hill-side bank and runs deep and green. Scenes like this one are familiar to everyone; we all know what a river is. We seldom consider that rivers are one of the most striking examples of fractal structures in nature. For our purposes, we define here the drainage basin of a river as the part of the territory where all the rainfall is collected by the same river and eventually transferred to the sea. A river network is composed of the collection of all the paths formed by every tributary of the main river in their drainage basin. This structure can be described as a graph where the vertices are the various parts of the landscape and the edges are given by the flow of water from one point to another. These various paths (tributaries and main streams) differ in the amount of water transferred and therefore the resulting structure is a directed weighted network. The size of the different parts of the system varies from hundreds of metres to thousands of kilometres (where this description starts to be inadequate) thereby spanning more than four orders of magnitude. Along these different scales and irrespective of the geographical locations, distinct river basins present a common structure.

The class of networks presented here in this chapter is somewhat different from usual graphs. A typical example of a river network is given in Fig. 7.1. The first thing we notice is that the graph is a tree. This happens because the force of gravity makes the water go only downhill. Therefore, we cannot have a directed path whose end-vertices coincide. Considering the network as non-directed does not change the situation. In this case we would have the possibility of cycle if a river stream divides at one point and reconnects downhill; but experimental evidence shows that rivers never divide and therefore no cycles are present even in this case. Everything is collected in a single point called the outlet, downhill in the drainage basin. It is natural to consider this point as the root of the tree. Secondly, a river network has the function of transferring the water for perhaps thousands of kilometres. It is then legitimate to think that such structure during its time evolution (on geological timescale) has been shaped in order to be as optimized as possible. In this case, this kind of optimization (if present) must result in some peculiar topological properties of this transportation tree. Thirdly, river networks are an example of networks embedded in a Euclidean space. This means they are a graph where the length and the size of an edge are important for the functionality of the system. The data used are extracted from satellite images where the pixels represent a zone of some hundreds of square metres. The edges connecting these vertices have a typical size and we cannot arrange as many as we want at the same point. It is therefore clear that we do not have to expect a scale-free distribution for the degree in this case. Nevertheless we will see how these apparently regular networks, develop unexpected and interesting long-range correlations resulting in different power-law distributions.

These points make the study of river networks particularly interesting. Here we can investigate how optimization of the transport properties results in particular graph topology. Application of these ideas ranges from food webs to the Internet. Different books (e.g. Rodríguez-Iturbe and Rinaldo 1996), and papers (e.g. Dodds and Rothman 2000a, 2000b, 2000c) are available for interested readers.

In this chapter we present a case study slightly different from the others. Here we introduce and describe the properties of river networks. Vertices of this graph are zones of the landscape and the edges are the path followed by raining water. Those networks are not scale-invariant in the sense of the degree distribution. Yet they show some self-similarity (in the more general form of self-affinity). The study of the scaling relations present in these aggregates can be of inspiration for the study of other networks. The properties of water transfer displayed by river networks (changing in shape and evolving along the whole history of Earth) can also help in understanding the onset of scale-invariance as a result of optimization processes in many other networks.

7.1   Satellite images and digital elevation models

All the territory of our planet is composed by different basins divided by their watersheds. Any of these drainage basins (the Amazon, the Nile, the Mississippi, etc.) contains in a hierarchical structure a series of nested (but not intersecting) smaller basins related to the various river tributaries.

By considering the amount of geological data collected in the past (Rodríguez-Iturbe and Rinaldo, 1996) it became evident that those structures present a self-similar behaviour. This is reflected in the appearance of power laws in the distribution of several quantities, chiefly total contributing area at a point (Rodríguez-Iturbe et $al.$, 1992a) and stream lengths (Tarboton, Bras, and Rodríguez-Iturbe 1988; Rinaldo et $al.$ 1996).

The network associated with a given river basin is experimentally analysed by using the digital elevation model (DEM) technique (Tarboton, Bras, and Rodríguez-Iturbe, 1988) which allows us to determine the average height of areas (pixels) of the order 10^{-2}km^2. The method is based on the phenomenological evidence that water falling from one site follows the steepest descent between those available. From one satellite photograph the zone is divided into portions whose linear size is about 20 - 30 m and each of these regions is treated as a single point. Then an average height z is assigned to these coarse-grained sites (from a satellite it is easy to compute the elevation field of a region). The network is built using the elevation field and starting from the highest mountain according to the steepest descent rule as shown in Fig. 7.2. Two pieces of experimental evidence are crucial at

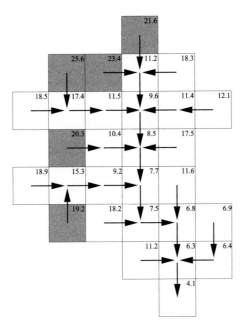

Fig. 7.2 An example of the DEM method on a sample elevation map. The mean height of the regions (reported in the upper right corner) are in arbitrary length units, for the sake of explanation. Grey cells correspond to peaks in the landscape, the network is built starting from them towards the outlet.

this point. The first is that a river never divides, therefore from any point there is one and only one path to the neighbours. The second that plays an essential role in the modelling of river network is that some portion of the territory may be unstable. In particular it has been noticed empirically that if A the mass of water in one point i and Δh_i is the slope of the steepest descent in the same point i it is always verified that

$$A^{0.5} \Delta h_i \leq constant. \tag{7.187}$$

The networks reconstructed in this way show a surprisingly universal self-affine nature (see Section 3.1.2). We can then say that river networks are a kind of geometrical fractals rather than true scale-free networks. The edges of these networks are real paths, made by rivulets and rivers with a characteristic width. Since one cannot accommodate hundreds of tributaries in the same point there is no scale-free distribution for the degree. Instead, the in-degree of the network (the graph is directed since water flows only downstream) is always somewhere between 1 and 3 (or $n-1$, where n is the number of nearest neighbours).

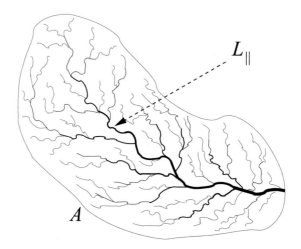

Fig. 7.3 Drainage (sub)basin and relative longest stream. The Hack's law states
$L_{||} \propto A^h$, $h \simeq 0.6$.

7.2   Geometrical scale invariance for river networks

As evident from DEM construction in these networks the in-degree can only
assume the values $1, 2, 3$ and the out-degree is always 1. Their frequency
distributions cannot be a power law. Nevertheless, many other topological
quantities are distributed in a scale-free shape. More like to fractal objects,
some of these scale-free quantities are related to metrical properties and
find their explanation either in the fractal shape of the landscape or in
a dynamical feedback of the type of self-organized criticality (see Section
4.2.1). We refer here to the scaling of the size of the basin with respect
to the length of the mainstream. Other quantities also have an unexpected
scaling as the distribution of stream lengths and the distribution of basin
areas. One of the main results in this field is that we can relate all these
different quantities by means of scaling relations.

Since the seminal work of J.T. Hack (1957) we known that the shape
of drainage river basins has particular geometrical properties. Distinct river
basins show the same self-affine behaviour (actually, fractals had not yet
been introduced in mathematics) valid on several orders of magnitudes. If
we take any traditional surface like a circle, a square, or a triangle, we note
that if the radius or the edge is doubled, then the area becomes four times
larger. Generally areas of compact objects grow as the square of the edge.
Reversing the relation we can also say that the edge of the structure grows

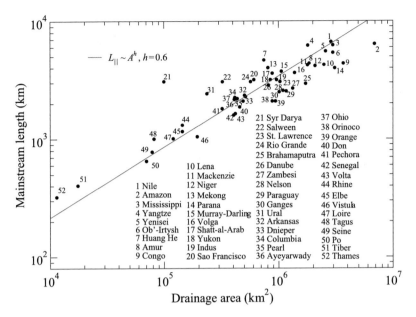

Fig. 7.4 Relation between lengths ($L_{||}$) and areas (A) of different rivers in the world. As a comparison we also plot (straight line) a fit with Hack's law.

like the square root of the area (by using formulas $L \propto A^{0.5}$). River basins are self-affine objects and they obey to the Hack's law shown in Fig. 7.3. This law states that,

$$L_{||} \propto A^{h}, \tag{7.188}$$

where we measure an exponent $h \simeq 0.6$.

The reason for that scaling is not related to the presence of empty regions as in a Sierpinski gasket. Rather, we have a difference between the geometrical properties along the mainstream (i.e. the length $L_{||}$) and in the direction orthogonal to the mainstream (i.e. the width L_o). This means that the two directions of the basin tend to grow with different regimes. Therefore in real basins there is a tendency toward elongation of the larger catchments. This means that basins tend to become longer and narrower as they enlarge (see Section 3.1.2). This behaviour was originally checked for the main basin and all sub-basins of Shenandoah River (Hack, 1965) but is actually quite general.[50] A plot with rivers of various size is shown in Fig. 7.4.

---

[50] Some attention must be paid on the fluctuations around this behaviour that can happen at small and large length scale (Dodds and Rothman, 2000a).

One of the first attempts to recover the fractal properties of mainstream distribution from basin shape is related to percolation theory (Green and Moore, 1982). The river is described as a directed percolation cluster ('lattice animal'). Making use of the theory of percolation, it is possible to predict this behaviour for the two lengths $L_{||}$ and L_o

$$L_{||} \simeq N^{\nu_{||}} \tag{7.189}$$

$$L_o \simeq N^{\nu_o} \tag{7.190}$$

where N (supposed to be very large) is the total number of bonds. Using N as a variable we can find a relation between the exponent of the area and those of the length.

☕ By noting that

$$A = L_{||}L_o \rightarrow A \simeq N^{\nu_{||}+\nu_o} \tag{7.191}$$

$$L_{||} \simeq N^{\nu_{||}} \tag{7.192}$$

we find that the Hack exponent is given by

$$h = \frac{\nu_{||}}{\nu_{||} + \nu_o}. \tag{7.193}$$

By using different techniques of computation for the two exponents $\nu_{||}$ and ν_o it can be found a numerical agreement with Hack's law. It has to be noted that if percolation is not directed, the two exponents are equal. This gives a 'natural' Hack's exponent of $1/2$. This could be the case for very large basins ($L_{||} > 10^4$Km) for which a deviation from Hack's law is reported. In this case the slope should be less relevant and the contribution of directed 'lattice animals' less important.

Starting from the geometrical shape of the basin we move to the quantities characterizing the network itself. The two most immediate quantities are the frequency $P(a)$ of basins whose size (i.e. number of pixels) is a (the size of a basin has been already introduced in the characterization of trees in Section 1.2). Similarly we can define the frequency $Q(l)$ of river streams whose length is l. As a consequence of the self-similar geometry of the basin the above quantities behave like a power law.[51]

$$P(a) \propto a^{-\tau}$$

$$Q(l) \propto l^{-\chi}. \tag{7.194}$$

The signature of fractal behaviour can be recognized by analysing different quantities, from the basin shape to mainstream length distribution or

---

[51] From this point A and $L_{||}$ refer to the area and length of the whole basin.

sub-basin area distribution. It is important to recognize that these different scalings are related to each other so that the various exponents are all correlated. This correlation can be sorted out by means of the *scaling relations*.

7.3   Scaling relations for river networks

7.3.1   Hurst exponent

If one considers the length $L_{||}$ of the longest stream in a certain area and the value A of this area, we already know

$$L_{||} \propto A^h. \tag{7.195}$$

Or in a different form (since the area is also by construction $A = L_{||}L_o$)

$$L_o \propto L_{||}^H. \tag{7.196}$$

This is the first scaling relation. It says that $H = \frac{1-h}{h}$ or more customarily

$$h = \frac{1}{1+H}. \tag{7.197}$$

Due to the similarity with the fractional Brownian motion, the exponent H is also called the *Hurst* exponent and all the other exponents are indicated from its value (Hurst, 1951).

7.3.2   Area distribution

Because of the presence of finite size effects, the real form of the eqn 7.194 is given by

$$P(a, L_{||}) = a^{-\tau} f\left(\frac{a}{A_c(L_{||})}\right)$$

$$Q(l, L_{||}) = l^{-\chi} g\left(\frac{l}{L_c(L_{||})}\right). \tag{7.198}$$

The f, g are scaling functions and $A_c(L_{||})$, $L_c(L_{||})$ are characteristic functions of the system length $L_{||}$ (Maritan *et al.*, 1996).

♣ In the case of areas $f(x)$ is a scaling function satisfying the following properties

$$\lim_{x \to \infty} f(x) = 0, \text{ sufficiently fast} \qquad (7.199)$$

$$\lim_{x \to 0} f(x) = c. \qquad (7.200)$$

As regards the exponent τ, we have $1 \geq \tau \leq 2$ because for $L_{||} \to \infty$ we want a power law for $P(a)$ with the requirement that $\int P(a)da = 1$.

Since we do not want any dependence upon $L_{||}$ in the argument of the function f and $A = L_{||}L_o = L_{||}^{1+H}$ we also have that the characteristic function $A_c(L_{||})$ must scale in this way

$$A_c(L_{||}) \propto L_{||}^{\phi}, \qquad (7.201)$$

where $\phi = 1 + H$.

By again using the normalization condition we obtain that

$$1 = \int_1^{\infty} a^{-\tau} f\left(\frac{a}{L_{||}^{\phi}}\right) da \qquad (7.202)$$

$$= L_{||}^{\phi(1-\tau)} \int_{L_{||}^{-\phi}}^{\infty} dx x^{-\tau} f(x) = \frac{c}{\tau - 1}. \qquad (7.203)$$

Therefore, one obtains that

$$P(1, L_{||} \to \infty) = c = \tau - 1, \qquad (7.204)$$

this means that *the number of sources in a basin is independent of the size of the system.*

By using the fact that rivers flow along a preferential direction we can expect that $\langle a \rangle \propto L_{||}$. Since by definition

$$\langle a \rangle = \int_1^{\infty} a P(a, L_{||}) da \propto L_{||}^{\phi(2-\tau)}. \qquad (7.205)$$

Then we find the first scaling relation

$$\phi = (2 - \tau)^{-1}. \qquad (7.206)$$

Since we have that $\phi = 1 + H$, we obtain the second scaling relation

$$\tau = \frac{(1 + 2H)}{1 + H}. \qquad (7.207)$$

Exponent	Scheidegger	OCN	Self-organized model	Real data
$h = \frac{1}{1+H}$	2/3	$0.57 - 0.58$	$0.65 - 0.72$	$0.57 - 0.60$
$\phi = 1 + H$	3/2	1.85 ± 0.05	1.38 ± 0.05	1.8 ± 0.1
$\tau = \frac{1+2H}{1+H}$	4/3	1.43 ± 0.02	1.43 ± 0.03	1.43 ± 0.02
$\chi = 1 + H$	3/2	1.80 ± 0.05	1.3 ± 0.3	1.8-1.9

Table 7.1 A table of the values for the various exponents characterizing the topology of a river basin

7.3.3   Length distribution

To study the length distribution, we need a change of variable that uses the 'convolution property'. Through this change of variable we can write

$$Q(l, L_{||}) = \int_1^\infty \Pi(l, a) P(a, L_{||}). \tag{7.208}$$

In this case we do not know either $Q(l, L_{||})$ or $\Pi(l, a)$, but we can guess some properties of the latter function. Since we know Hack's law, this function must not very broad, but rather peaked as a function of one variable with respect to the other. In particular we can use the mathematical form of Hack's relation into the delta function $\Pi(l, a) = \delta(l - a^h)$.

 At this point the above equation becomes

$$Q(l, L_{||}) = \int_1^\infty \delta(l - a^h) P(a, L_{||}). \tag{7.209}$$

This equation can be evaluated in the limit $L_{||} \to \infty$ giving that

$$Q(l) = l^{-1-\frac{\tau-1}{h}}. \tag{7.210}$$

We then obtain the third and latest scaling relation

$$\chi = 1 + \frac{\tau - 1}{h} = 1 + H. \tag{7.211}$$

In such a way the universality value of h reflects in the universality of the value of H, τ, χ for every river on Earth.

7.4   River networks models

7.4.1   Scheidegger model

The first and most immediate model of a river network works by random aggregation of streams. Consider that the landscape is a simple incline and

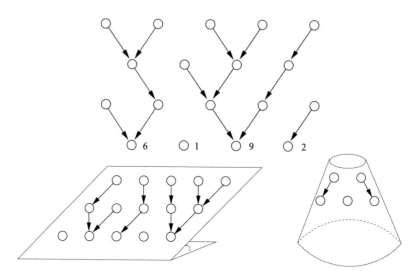

Fig. 7.5 Above: an example of the aggregation of streams. Below: the landscape considered by the model, with fixed and periodic boundary conditions. Note that as the number of vertices in the layer grows, the two different conditions give the same results.

the vertices in the networks are arranged according to a tilted square lattice (see the upper part of Fig. 7.5). In this way any vertex has two immediate neighbours downstream. Every vertex in one layer of the incline can forward the water with equal probability either to the vertex below on its right or to the vertex below on its left. Vertices on the boundary of the incline can transfer water only to one of the two destinations below. Usually they are connected through periodic boundary conditions (see the lower part of Fig. 7.5) (Dodds and Rothman, 2000a). For simplicity, we also take the size of the lattice to be 1. In this way, this model corresponds to the model of the directed random walk (Feller, 1968). Therefore, the path followed by water is a directed random walk and the size of a basin can be defined by considering the boundaries to be made by random walks. One basin starts with two contiguous random walks and it terminates when the walks cross each other (see Fig. 7.6). Size and length of a basin are then obtained by knowing the probability of 'first return' for a directed random walk. The number of steps l necessary for two walkers originating in the same place to cross again is distributed according to

$$P(l) = \frac{1}{2\sqrt{\pi}} l^{-\frac{3}{2}}.$$

(7.212)

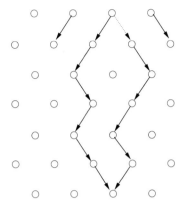

Fig. 7.6 An example of a basin in a Scheidegger model. The two boundaries are formed by random walks. Two walkers start from neighbouring sites, when they cross in the same horizontal position they form the outlet. We make the approximation of consider the right-hand walker as starting in the same point (indicated by dotted arrow, the difference vanishes for large basins).

Since this number n is proportional to the length of the basin l we find the first exponent for the length distribution $\chi = \frac{3}{2}$. We are now in a position to derive through scaling relations all the other exponents (they are listed in Table 7.1). Note that h can be found self-consistently under the random walk picture. This is because the typical area of basin is given by $l \times l^{1/2}$ where the transverse dimension is the wandering of a random walk. this gives $a \propto l^{3/2} \propto l^{1/h}$, from which $h = 2/3$.

7.4.2 Optimal channel networks

A way to model the time evolution of river network structures is given by the optimal channel networks (OCN). The networks produced by this model are the result of an evolutionary process. This means that several different configurations are tested and discarded. The way to select one configuration with respect to another is to measure a specific quantity that is the total loss of gravitational energy in the system. This corresponds to assuming that on Earth, the driving force shaping the landscapes and the structure of drainage basins is the search of the minimum for such an 'energy'.

Gravitational energy is a way to take into account the outcome of the work done against the gravitational field. To illustrate, if we carry a weight to the top floor of a building our efforts are transformed into increased gravitational energy of the weight. This energy can be recovered as kinetic energy by dropping the weight out of the window (do not try that at home). The analytical form of the gravitational energy for a weight of mass m is given by $mg\Delta h$, where Δh is the difference in the position of the body (before and after the work) and g gives the strength of the gravitational energy. In principle, g is slightly variable on Earth's surface. However, it can be considered constant within a drainage basin. The total loss of gravitational energy is then

$$G = \sum_{i=1}^{L^2} M_i g \Delta h_i. \tag{7.213}$$

With an opportune change of unit, this expression can be written as

$$G = \sum_{i=1}^{L^2} a_i^{1/2}. \tag{7.214}$$

This is because the mass of water collected in a point is given by the sum of masses transferred from upstream zones and then is proportional to the area of the basin upstream. This gives $M_i = ka_i$. The eqn 7.213 then becomes

$$G = gk \sum_{i=1}^{L^2} a_i \Delta h_i. \tag{7.215}$$

At this point we make use of experimental evidence shown in eqn 7.187, that is the gradient Δh_i on site i is known to behave

$$\Delta h_i \propto a_i^{-0.5}. \tag{7.216}$$

One can then write eqn 7.215 as

$$G = \sum_{i=1}^{L^2} a_i^{1/2} \tag{7.217}$$

where, with an opportune choice, we removed the constant kg from the definition of G.

The value of G is different for any realization of the basin. The model starts from one basin with a certain value of G then produces slightly different basins with different values of G and accept the variation if the value of G decreases. OCNs are defined as the class of networks for which G is the minimum possible. The numerical and analytical results obtained by studying this model are in very good agreement with experimental data.

This allows us to divide the various networks into different families by considering under which circumstances the following function

$$H = \sum_{i+1}^{L^2} a_i^\gamma \qquad (7.218)$$

assumes a minimum value (Rinaldo *et al.*, 1996).

When $\gamma < 0$ then the mainstream passes for all the points of the landscape, in the language of networks this corresponds to having *Hamiltonian paths* for the rivers (Huggins, 1942; Nagle, 1974). In the case $\gamma > 1$ there is as little aggregation as possible. These paths are called *explosion paths*. The most interesting situation happens when $0 < \gamma < 1$. This case can be explored by computer simulations. In particular one can define an evolution in the space of the possible networks. One starts with an initial condition and changes one branch of the tree, this change is accepted with a probability proportional to $e^{-\Delta E/T}$. In this case T is a parameter that is gradually tuned to 0. From an initial condition this 'random' spanning tree evolves to a river-like structure.

7.4.3   Self-organized models

Landscape evolution models describe the dynamics acting in the drainage basins (Howard, 1994). The river network is a two-dimensional projection of the three-dimensional tree-like structure given by computing the steepest descent on a rugged surface.

Based on this starting point some self-organized models have been introduced to reproduce the observed data (Rinaldo *et al.*, 1993; Caldarelli *et al.*, 1997). Starting from a three-dimensional landscape we again have an evolutionary process in which different states are explored by the model. The rules are very similar to the ones proposed in self-organized critical (SOC) models (see Section 4.2.1) of sandpiles (Bak, Tang, and Wiesenfeld, 1987). Whenever the local shear stress exceeds a given threshold, erosion starts an 'avalanche', the landscape changes and a related rearrangement of the network patterns takes place. In its simplest formulation the landscape is described by a lattice. For every site i of a square lattice of size $L \times L$ we define a height h_i. The lattice is then tilted at an angle θ (with respect to a given axis) to create a gradient. Usually two possibilities are analysed: All the sites on the lowest side (kept at height $h = 0$) are possible outlets (i.e. the multiple outlet arrangement) of an ensemble of rivers which are competing to drain the whole $L \times L$ basin. Only one site is kept at $h = 0$ and it is the outlet of a single river in the $L \times L$ basin. In addition, for both the above cases, two types of initial conditions are usually considered

(Caldarelli, 2001): (a) a regular initial landscape, for example, flat, and (b) an irregular surface obtained by superimposing a suitable noise on a smooth sloping surface. Each site collects a unit amount of water from a distributed injection (here a constant rainfall rate as in the original approach) in addition to the flow which drains into it from the upstream sites. A unit of water mass is assigned to each pixel in the landscape. Therefore, the total area drained into a site is also a measure of the total water mass collected at that site. From each site water flows to one of eight sites, four nearest neighbours and four next nearest neighbours along the steepest descent path.

The time evolution of the model proceeds according to the following steps:

1 The shear stress τ_i acting at every site is computed as (Rinaldo *et al.*, 1993)

$$\tau_i = \Delta h_i \sqrt{a_i} \qquad (7.219)$$

where Δh_i is the local gradient along the drainage direction.

2 If the shear stress at a site exceeds the threshold value, τ_c, then the corresponding height h_i is reduced (e.g. by erosion) in order to decrease the local gradient. The shear stress is then set just at the threshold value. This produces a rearrangement of the network followed by a re-updating of the whole pattern as in step 1.

3 When all sites have shear stress below threshold, the system is in a dynamically steady state. Since this situation is not necessarily the most stable, a perturbation is applied to the network with the aim of increasing the stability of a new steady state. A site is thus chosen at random and its height is increased in such a way that no lakes, that is sites whose height is lower than that of their eight neighbours, are formed. Steps 1 and 2 then follow as before.

After a suitable number of the perturbations (step 3), the system reaches a steady state which is insensitive to further perturbations and where all statistics of the networks are stable. This resulting state is scale-free, that is, it is characterized by power-law distributions of the physical quantities of interest.

The largest numerical calculations have been carried out on a bidimensional square lattice (where each site has eight nearest neighbours) for sizes up to $L = 200$ (Caldarelli, 2001) with reflecting boundary conditions in the direction transverse to the flow and open boundary condition in the parallel one.

Very interestingly, during the time evolution of the landscape it is possible to check the total energy dissipation of the system, previously defined as $G = \sum_{i=1}^{L^2} a^{0.5}$ (where i spans all sites of the lattice) (Rodríguez-Iturbe

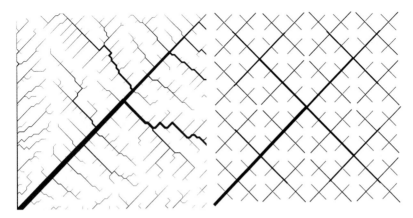

Fig. 7.7 Two kinds of optimization. Left: a 'feasible' configuration. Right: the absolute minimum of total gravitational energy dissipation.

et al., 1992*b*). It is interesting to note that in this model where no hypothesis is made on G, we still observe an almost monotonic decrease of G. The structures obtained are very similar to that of OCN. In Fig. 7.7 we see a computer evolution of a river network (left) and the absolute minimum represented by a structure known as Peano curve (right).

7.4.4   Allometric relations

A way to quantify such transport efficiency of a tree-like structure is given by the *allometric relations*, $C(A)$. The general derivation is given in Section 5.7.1, here we only report the little part related to river networks. Given the various areas a_i in a basin whose area is A, we have that $C(A)$ for the basin A is defined as

$$C(A) = \sum_{i \in A} a_i \qquad (7.220)$$

where i is summed over all the vertices in the sub-basin A.

To understand the physical meaning of this quantity in the case of river networks one has to imagine that the water flows uphill from the outlet to the various sources. Then the value of $C(A)$ is the total supply of water that is needed in order to feed the various points in a basin of area A. If this supply scaled (i.e. growing or becoming smaller) like the area, then the relation between the two would be 'isometric'. Instead in a variety of physical systems from river networks to blood and lymphatic systems the scaling is of the kind $C(A) \simeq A^\eta$ where the exponent $\eta \neq 1$ and the relation is therefore 'allometric'. Not all the choices of the exponent of course result in a different supply 'shape'.

The smaller the scaling of $C(A)$ with A, the more efficient the transportation system. Banavar, Maritan, and Rinaldo (1999) provide a proof that, for a tree-like transportation system embedded in a D-dimensional metric space, the most efficient scaling of $C(A)$ as a function of A is given by $C \propto A^{\eta}$ where $\eta = \frac{D+1}{D}$. This correctly predicts the observed scaling in river networks ($\eta = 3/2$, $D = 2$) and in respiratory and vascular systems ($\eta = 4/3$, $D = 3$) (See also West, Brown, and Enquist 1999). In general, if the tree-like network is not embedded in any metrical space (such as spanning trees derived from food webs or from the Internet), any vertex can be connected with any other, this results in an effective dimension tending to infinity, which gives $\eta = 1$.

7.5   River networks on Mars' surface

As an application of river networks we can now use the universal behaviour of eqn 7.194 to predict if erosion took place in some other environment; in particular the landscape of Mars. Presence of water on the planet would increase the probability of some past form of life. Presently, the atmospheric conditions are almost incompatible with surface liquid water, so that its presence in the past has to be detected indirectly, looking at clues in the geomorphology of the planet. The possible presence of drainage basins on Mars traces back to the Viking mission in 1977. At the time this mission returned the picture of an area known as Warego valley where the presence of a typical erosion pattern is evident (see Fig. 7.8 left). Recently, the high-resolution pictures taken in the last three years by the Mars Orbiter Camera (MOC) camera of the NASA probe Mars Global Surveyor (MGS) have provided some new evidence (Malin and Edgett, 2000a) of the presence of liquid water in (geologically) recent times. Other traces can reconnect to the presence of lakes (Malin and Edgett, 2000b) and oceans (Head III, Hiesinger, Ivanov, Kreslavsky, Pratt, and Thomson, 1999).

All this photographic evidence must be complemented by some more robust analysis on data such as the ones from the Mars Orbiter Laser Altimeter (MOLA).[52] Through MOLA it is possible to show that the channel networks of some parts of the planet are very similar to the structures carved by water

---

[52] The MOLA is a satellite orbiting the planet. It sends a laser beam to the planet surface. Measuring the time elapsed when the beam comes back and knowing the height of the orbit, it is possible to obtain a measure of the landscape altitude. Measures are very dense along the projection of the orbit on the surface, less dense in the orthogonal direction. by collecting data on the several orbits it is possible to divide the planet surface into zones and assign an average height to them.

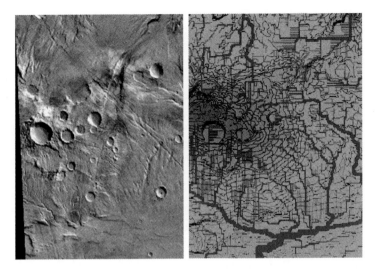

Fig. 7.8 A picture of a Martian valley and its DEM computed from MOLA data.

erosion on Earth. A preliminary analysis was first reported in Caldarelli, De Los Rios, and Montuori (2001) and lately in Stepinski *et al.* (2002), and in Caldarelli *et al.* (2004c).

The latest MOLA data provide a DEM such that every pixel covers a surface of about 1690×1690 m² (roughly $0.03125° \times 0.03125°$). Even if data with a larger resolution have not been released yet, it is possible through MOLA software to extract information on a grid where any pixel covers an area of 845×845 m². With these data at hand, it is possible to select four regions of about 100×100 km² and reconstruct a network from the DEM (Caldarelli, De Los Rios, and Montuori 2001; Caldarelli *et al.* 2004c). These regions are the *Warrego Vallis* shown in Fig. 7.8 (left), the *Solis Planum* and *Noctis Labyrinthum*, two rough districts in the Tharsis region, and a region from the northern hemisphere where an ocean could have existed in the past (Head III, Hiesinger, Ivanov, Kreslavsky, Pratt, and Thomson, 1999). The results of this reconstruction are shown in Fig. 7.8 (right), where as usual every part of the network has a width related to the area it collects. In Fig. 7.9 there is statistical analysis of the basin areas of the Warrego Vallis. To interpret the results it is important to understand the dependence of the analysis upon the application of the 'pit removal' procedure.

Pit removal is a way to take into account the fact that streams can sometimes fill local depressions, forming lakes. Once the water fills the difference in height between the depression and the first neighbours the stream continues its path to the outlet. On rough terrains the pit removal procedure is extremely important since before its application the drainage basin would be

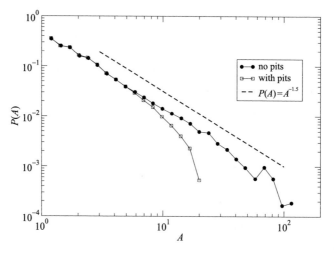

Fig. 7.9 The plot of $P(A)$ for the region shown in the previous image of Warego Valley.

fragmented into very small sub-basins where pits act as outlets. Only after pit removal does the whole river network emerge. The $P(A)$ distribution in Fig. 7.9 is shown both before and after pit removal. Before removing pits the distribution seemingly exhibits a power-law decay with exponent τ close to 1.5. The same result holds for the other terrains, Solis Planum and Noctis Labyrinthum (not shown), whereas the northern hemisphere areas decrease exponentially. These first results suggest that the northern lowlands are extremely flat, without any long-range correlated structure. On the contrary, the three other regions show some correlation, at least for small areas. The corresponding network for the Warrego Vallis is shown in Fig. 7.8 (right). only increases the cut-off so that the exponent remains 1.5 Once pits are removed we find a behaviour similar to that of the Earth.

The exponent γ cannot be fitted before pit removal, whereas it is close to 2 after pit removal, consistent with the relation $\tau = 2 - 1/\gamma$. It is important to stress that the exponent $\tau = 1.5$ cannot be simply the result for a network reconstructed on a random slope. Both numerically and analytically it has been found that in this case $\tau = 4/3$ (Scheidegger, 1970). Therefore the exponent $\tau = 1.5$ is different from both flat and sloped random networks.

8. Ecological networks

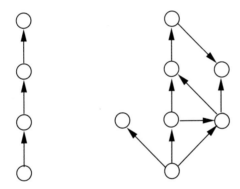

Fig. 8.1 Left: an example of a food chain. Right: a slightly more complicated food web with some interconnected food chains.

One hot spring evening, just as the sun was going down, two men appeared at Little Rock Lake. Their work was long and difficult, they had to look for all the species present in the lake and the predation relationship between them. For almost 200 species this work could last years. This work eventually resulted in the publication of one of the largest data sets available (Martinez, 1991). For every species one has also to discover the predation relationship with all the others. It was very tiring but necessary work. Like any description of an ecosystem, any model must include the number of species and the predators and prey present in the area. This structure is naturally described by a network whose vertices are the species and whose directed edges are predations.[53] In the ecological literature those systems are called food webs.

One of the first examples of a food web to be studied was the very simple topology of a food chain (Elton, 1927). The idea behind this quantity is that every species must take the resources to survive from the abiotic environment. Plants or bacteria are the first to transform light, minerals, water, and sunlight. A successive series of predations transfers these resources to the top predators as shown on the left of Fig. 8.1. The concept of a food chain reproduces the fundamental fact that every species is connected by a path of successive predation to the environment. This chain is the ecological counterpart of the topological distance in a graph. Indeed, a well-known fact in ecology is that the length of these food chains is often limited to five, six steps. This experimental result is evidence of the small-world effect in ecology (even if the origin is more complicated than a shortcut effect). A more realistic picture of a natural environment is given by considering a series of interconnected chains, for example the structure shown in Fig. 8.2. In this complicated topology, it is a formidable task to predict the effects of modification in the food resources.

The stability of food webs is not only a scientific problem but an issue also in economics and politics (Buchanan, 2002). Problems like the exploitation of food sources are becoming more and more frequent. The Peruvian anchovy fishery collapsed in 1971 and twenty years later a similar phenomenon repeated for the North Atlantic cod fishery (Pauly et al., 1998). After times of massive indiscriminate fishing the result has been the dramatic reduction of predators. Fishery then moved to more basal species like anchovies, but these species also started to decrease dramatically. When big predators like tuna or cod (predators) are removed from the web, other predators rule the system in absence of competition. The increase of these non-edible fishes completely depletes what is left of commercial value. Since the various connections between species are completely intertwined it is very difficult to predict any change in the composition of the system. For example, cod predators are also predators of cod competitors. Nobody knows if a reduction in those predators results in a net increase in the cod population.

It is clear that for all the above reasons any progress in the field of food webs can be potentially very important. It is not surprising then, that the first studies of population dynamics are at least as old as the findings of Fibonacci in the thirteenth century (for the increase in rabbit

---

[53] In order to describe the system dynamics properly, it would be important to know also the amount of the predation (this would allow us to deal with a weighted network). Unfortunately this piece of information requires an extremely large amount of field work, therefore many studies of food webs do not report those data.

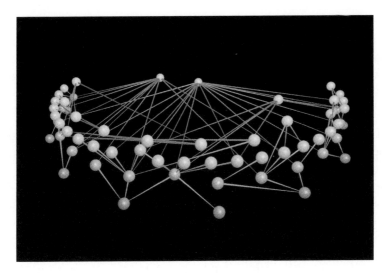

Fig. 8.2 A simple food web of a grass-based ecosystem in UK. Courtesy of N. Martinez.

populations). Studies on population dynamics increased since then and have been the subject of the activity of scientists like Alfred James Lotka[54] and Vito Volterra.[55] The way to describe the process of population dynamics is through the equations giving the growth rate g (the difference between the birth rate b and the death rate d). A complete description of the models of population dynamics is given in Appendix E. Starting from these approaches, modern graph theory tries to characterize these systems and predict their behaviour.

Similar problems are also present for plant communities that compete for the available resources. One possible approach to determine species interaction (there is no predation) is to study their similarity. There are several criteria that can be used to measure such a similarity. The traditional one was introduced by Carolus Linnaeus who described the flora and fauna of European ecosystems. This classification scheme is now used worldwide, characterizing all the living organisms at a basic level called 'species'. Different species are then grouped in a category (called taxon) called 'genus'. The various categories (taxa) are recursively grouped, in a hierarchy of clustering. Eventually the categories are so general that there is one for plants

---

[54] Statistician and ecologist born in Lemberg Austria (now Ukraine) in 1880 and died in 1949 in USA.

[55] Italian mathematician born in Ancona, Papal States (now Italy) in 1860 and died in Rome in 1940.

and another for animals (this taxon is named 'kingdom'). The result of this process can be viewed as a tree graph. From the more general taxon that corresponds to the root (i.e. plants) we have various offspring at branching levels. Also in this latter case the study of the topology of this tree reveals some unexpected features.

> In this chapter we present some ideas behind the study of ecological systems. As regards animals, traditional theory focuses on the population dynamics of the various species. Very briefly we give some indication of the traditional models of population dynamics. In any approach the key quantity is knowledge of the predations, i.e. the edges that define the food webs. In the case of plants (where predation is not present), we still have a correlation between species. This correlation presents remarkably self-similar properties, as can be found by considering the tree obtained by classification.

8.1　Species and evolution

The modern theory of species diversification traces back to the work of Charles Darwin.[56] The idea is that morphologically every species is characterized by a series of features (lately known to be determined by the genetic code). At every individual generation this set of features can mutate randomly, that is to say new features can be added and/or some of the old can be removed from the species. Some mutations are dangerous, in the sense that the individual cannot live with these new features (e.g. a different composition of the blood). Some others are not dangerous and the individual can live with those. In principle we cannot say if a mutation of one kind is 'better' than another. But in practice, some mutations can produce new features that give an advantage to some individuals with respect to other individuals without the same feature. The environment (i.e. other individual, species, and resources) determines a selective pressure on the corpus of mutations by selecting some of them and discarding others. Ultimately, the environment determines the evolution of the species. Note that this is a feedback process, since the environment is not only a set of fixed external conditions, but is also transformed by the species present in it. This is the

---

[56] British naturalist born in 1812 at Shrewsbury, England and died in 1882. His ideas and theories had a tremendous impact not only on his field but on society as a whole.

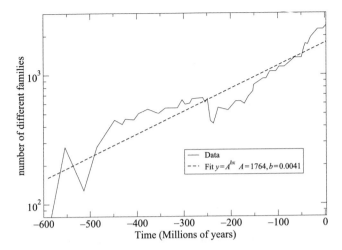

Fig. 8.3 The increase of diversity in the Phanerozoic era. After Benton (1993).

reason why many authors prefer to refer to the concept of 'co-evolution' of species, where now the external environment is a complicated series of constraints given also by the presence of other individuals of the same or different species.

A way to measure the relation between species and environment is given by the concept of species *fitness* (Kauffman, 1993). Large fitness values for one species represent a measure of the species success. In order to be precise we must restrict the definition of fitness to a single phenotypic character rather than to individuals. In the former case we can then say that *the fitness of a certain character is given by the average number of descendants produced by an individual with that character.* It is very difficult to be equally precise if we want to define fitness for a species or for a group of individuals (this ambiguity is related to the difficulty in defining the concept of species). Generally (especially in simplified models) one means by the fitness of a species the fitness of a phenotype. This means the average number of descendants of the individuals with a specific phenotype. Note that this concept could be misleading, since fitness defined in this way (number of descendants) is generally dependent upon various conditions external to the phenotype.

8.1.1 Mutations, extinctions population dynamics

In the evolution of an ecosystem we have two different dynamics. On 'short' timescales (life of individuals) we have an increase or decrease in populations for a fixed set of species. On a longer timescale, new species can appear and

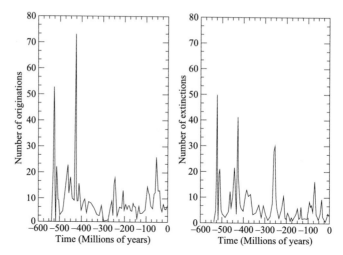

Fig. 8.4 Left: the number of extinctions in the Phanerozoic era. Right: the number of originations. Courtesy of M.E.J. Newman (Newman and Sibani, 1999).

others disappear (when the decrease of the individual number becomes small enough). This is common in the history of the Earth, and from the fossil record we have a clue of the rate of species creation and destruction. Some of the main data sets analysed by scientists refer to only one period, the 'Phanerozoic' era (Benton 1993; Sepkoski Jr 1992, 1993). This geological era covers the last 544 million years of life on this planet. This period is traditionally divided into 77 sub-periods of about 7 million years each. By considering how the species appeared and disappeared in the various sub-periods we can have a measure of the ecosystem evolution. There are various problems with these data sets. Firstly the real duration of the period is not known, secondly the various sub-periods are not of the same duration. Another point is that very old fossils tend to be less well preserved than more recent ones, giving a total bias on the set of measures. Finally, we do not have any control on the extinction of species for which we did not find any experimental evidence.

As shown in Fig. 8.3 the total diversity can be measured by considering the increment with time in the number of different families. This number grows fairly smoothly even if it is the result of a very complicated dynamics. This can be shown by considering the number of extinctions (see Fig. 8.4) for which we find a less smooth behaviour. We see that the dynamics of evolution is extremely discontinuous, with big events where many different species disappear simultaneously from the environment. We have experimental evidence of five great extinctions: the Ordovician, Devonian, Permian, Triassic, and Cretaceous. In the case of the Permian, the extinctions involved more than 90 per cent of existing species.

In most cases, special events, such as meteorites, volcanic and seismic activity, or change in the climate, can be considered as the primary causes of these extinctions. Others believe that those mass extinctions might be caused by small external perturbations. To capture this behaviour, the idea of *punctuated equilibrium* was proposed (Gould and Eldredge, 1977). Species undergo rapid mutations and ramification from a parent species in a relatively short period. For the rest of the time species remain static. Species dynamics is then composed of many stasis period with a few periods of fast change. Conversely, a few extinctions in the system can produce a chain reaction eventually destroying almost all the species in the ecosystem.

8.2   Food webs: a very particular case of network

Whatever the model considered, for larger or shorter timescales, an essential ingredient is given by species interaction, i.e. predations. If species are vertices and the predations are the edges we obtain naturally a food web. Since predations are not symmetric, the food web is a directed graph. We follow the convention (based on the flux of nutrients) of drawing the arrows from the prey towards the predator. Food webs have been studied for a very long time and ecologists have defined a series of biological quantities that have an interesting counterpart in the topological properties of food webs.

At the base of the system there are plants and simple organisms like bacteria. They are able to convert water, minerals and sunlight into living material that is used by predator species to sustain themselves. Typically food webs have a small size, on average we have about 50 to 60 species. Both the order (number of vertices) and size (number of edges) is much smaller than respect to all the other case studies presented in this book. In food webs vertices are not at all the same. Some vertices (primary producers) are sustained by sunlight or other sources in the external environment. Others must predate to live. This solution is not at all satisfactory from the point of view of the amount of resources obtained, since in the steady state, predators can count on only 10 per cent of the prey. This issue is at the hearth of the small-world properties of food webs. Ecologists use to divide the species into three different classes: *top*, *basal*, and *intermediate* species. Respectively, these classes contain species with no predators, species that do not predate, and all the species that are both predators and prey, as shown in Fig. 8.5. The number of predations between different species *basal-intermediate (BI)* *intermediate-intermediate (II)*, *intermediate-top (IT)*, and *basal-top (BT)*

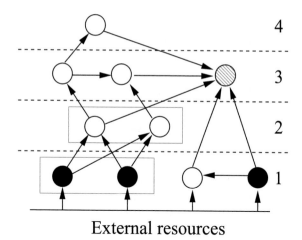

Fig. 8.5 A simplified food web. The black species are basal species (no prey); the grey ones are top species (no predator); all the rest are intermediate. The level of one species is given by the shortest number of paths separating it from the external resources. As in this case not all the species on the first level are necessarily basal. Similarly, not all the species on upper level are necessarily top ones. Rectangles represent the trophic species (those with same predators and same prey) that can be defined in the web.

are also typical quantities of interest (Cohen, Briand, and Newman, 1990; Pimm, 1991).

During field observation it is very easy to distinguish between an eagle and a hawk. However, one can cluster several different species of bacteria in the same 'species'. This overestimates the proportion of predators with respect to the prey. Usually to avoid such bias the concept of *trophic species* is used. A trophic species consists of all the species that predate on and are predated by the same species. This is an important point, since in food webs we have one of the few cases where we can cluster together different vertices as shown in Fig. 8.5 (see Section 2.3).

8.3   Food web quantities

As already mentioned, few food webs data are available. We list below some quantities to characterize them and we report the corresponding features in Table 8.1.

Food web	N	χ	D
Skipwith Pound (Warren, 1989)	25	0.31	1.33
Coachella Valley (Polis, 1991)	29	0.31	1.42
St. Martin Island (Goldwasser et al., 1997)	42	0.12	1.88
St. Mark Seagrass (Christian et al., 1999)	48	0.10	2.04
Grasslands (Martinez et al., 1999)	63	0.02	3.74
Silwood Park (Memmott et al., 2000)	81	0.03	3.11
Ythan Estuary 1 (Hall and Raffaelli, 1991)	81	0.06	2.20
Little Rock Lake (Martinez, 1991)	93	0.12	1.89
Ythan Estuary 2 (Huxham et al., 1996)	123	0.04	2.34

Table 8.1 Properties of some food webs. N represents the number of trophic species, the value χ is the connectance, D is the diameter.

8.3.1 Connectance

The connectance is a quantity related to the topology of the web. It represents the number m of edges (predations) present in the food web with respect to the possible ones. In formulas (remember that this graph is directed so that the total number of edges is $n(n-1)$):

$$\chi = \frac{m}{n(n-1)} \simeq \frac{m}{n^2}. \tag{8.221}$$

This quantity is the analogous (apart a factor 2) of the density of links ρ defined in eqn 5.178. It has been considered in a series of different studies, even if the results are somewhat controversial. Historically, some studies (Sugihara, Schoenly, and Trombla, 1989; Pimm, Lawton, and Cohen, 1991) seem to discover a monotonic increase of the type $\chi \propto n^\alpha$ for the connectance χ with respect with the graph order n. More recent studies (Martinez, 1992) found instead an almost constant behaviour for a series of tropical ecosystems. Some new analyses deviate from this 'constant connectance hypothesis'. In this book we shall assume that at the moment no clear trend is understood for connectance in different food webs.

8.3.2 Composition of food webs

Some attention has been paid to the question of whether top, basal, and intermediate species show a typical proportion everywhere or are relative to the particular ecosystem considered. A traditional point of view on the basis of the data collected at the time (Cohen, 1989; Cohen, Briand, and Newman, 1990) showed that a specific proportion for top (T), basal, (B) and intermediate (I) species was present. In particular data seem to show an almost

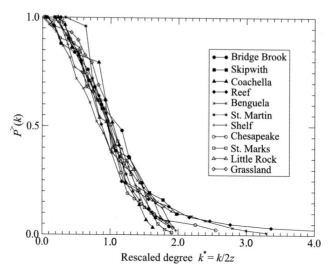

Fig. 8.6 The plot of the cumulative degree distribution. For no data set is there a clear power-law tail. When all the degrees are normalized by dividing them for the ratio $z = L/N$ of edges and vertices in the web, a common behaviour appears. Courtesy of D.B. Stouffer (Stouffer *et al.*, 2005).

universal behaviour across different webs (Briand and Cohen, 1984; Havens, 1992) where it a proportion of $T \simeq 25\%$, $B \simeq 40\%$, $I \simeq 35\%$ was found. However, in this case the most recent analysis (Martinez, 1991; Martinez and Lawton, 1995; Goldwasser and Roughgarden, 1997; Caldarelli, Higgs, and McKane, 1998) shows a different result. Regardless of the particular environment, one might expect that the majority of species is intermediate, consequently almost all the edges are of type II. A possible explanation for such discordance could be the poor quality of observations. This means that most of the predations could be lost during the data collection. This can be confirmed by randomly removing the food web edges (in order to mimic non-observed predations). This effect is known to alter the food web properties substantially (Goldwasser and Roughgarden, 1997). This idea has also been tested for a model of food webs (Caldarelli, Higgs, and McKane, 1998) where every predation relationship is under control; in this case every single predation is known from computer simulation. For almost any value of the parameters of the model one finds a vast majority of intermediate species, but as the predations are gradually randomly removed we recover the proportion of 25 per cent of top species.

8.3.3   Degree distribution

Some studies (Montoya and Solé, 2002) found a power-law degree distribution with an unusually small value for the exponent ($\gamma \simeq 1.1$). In this case the finite size effect related to the small size of these webs has a strong effect on the distribution $P(k)$. An extensive report (Stouffer $et\ al.$, 2005) found for the data shown in Fig. 8.6 a fit of the kind

$$P^>(k^*) = e^{-k^*} - k^* \int_{k^*}^{\infty} \frac{e^{-k'}}{k'} dk', \qquad (8.222)$$

where k^* denotes the value of the degree k divided by the quantity $2z$, where $z = m/n$ is the ratio of edges m with respect to the number of vertices n. In this way degrees relatives to different data sets (with very different values of z) can be compared.

As regards the robustness of ecosystems, food webs display a great stability under species removal. Avalanches of extinctions do not happen when removing any of the species in the data recorded. This is particularly remarkable since the graph connectivity is not dominated by the species with the largest degree (i.e. the largest number of connections). Therefore the ability to develop different habits of predation ensures the stability of the whole system.

8.3.4   Distance

The notion of distance is rather important for food webs. At any predation neither all the individuals in prey species, nor all the biomass of individuals predated is transferred to the predator. Therefore the larger the number of consecutive predations to reach a top species, the lower the resources effectively transferred. Generally then, food webs are characterized by a small diameter (Williams $et\ al.$ 2002; Montoya and Solé 2002; Camacho, Guimerà, and Amaral 2002).

8.3.5   Trees in food webs

It is possible to discover a scale-invariance of the system by reducing the complexity of the food web to that of a spanning tree (Garlaschelli, Caldarelli, and Pietronero, 2003). A natural way to find a spanning tree is given by selecting the strongest predation relation for any species. Also in this case the degree distribution does not change its shape. Once the food web is transformed in a tree, it is possible to study how the resources are delivered in the ecosystem. To that purpose, we make use of the concepts on allometric relations described in Section 7.4.4. For the food webs it is also possible to recover a scaling of the kind $C(A) \propto A^\eta$. Interestingly, as

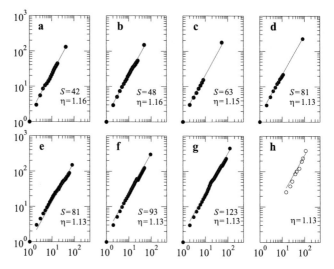

Fig. 8.7 The scaling of $C(A)$ against A for various food webs with the best power-law fit. **a** St Martin Island, **b** St Marks Seagrass, **c** grasslands, **d** Silwood Park, **e** Ythan Estuary without parasites, **f** Little Rock Lake, **g** Ythan Estuary with parasites, **h** the plot of $C(A_0)$ against the area A_0 at the root of the tree for a collection of webs including those in **a-g**. After Garlaschelli, Caldarelli, and Pietronero (2003).

shown in Fig. 8.7 the value of the exponent $\eta = 1.13 \pm 0.03$ is the same within the error in almost all the data sets available. This behaviour is further confirmed by plotting the allometric scaling $C(A_0)$ for different data sets represented by the area A_0 at the root vertex (corresponding to abiotic environment). A value of the exponent $\eta = 2$ signals the least optimized distribution, while optimized trees embedded in a Euclidean dimension d have an exponent $\eta = \frac{d+1}{d}$ (see Fig. 4.6). These experimental results show that the value of the exponent η is smaller in food webs than in other systems. This is due to the absence of any Euclidean dimension d in this case; nevertheless, the spanning tree of a food web does not reach the absolute minimum $\eta = 1$ given by a star-like structure. The measured value of η is a compromise between the minimization and the competition between species. For the latter reason species arrange themselves in different trophic layers and avoid to predate on the same resource (Garlaschelli, Caldarelli, and Pietronero, 2003).

Kingdom	Plants	Plants	Plants	
Phylum	Anthophytae	Anthophytae	Anthophytae	
Class	Dicotyledones	Dicotyledones	Liliopsidae	
Order	Solanales	Solanales	Musales	
Family	Solanaceae	Solanaceae	Musaceae	
Genus	Lycopersicon	Solanum	Musa	
Species	Lycopersicon esculentum	Solanum melongena	Musa paradisiaca	
	TOMATO	**AUBERGINE**	**BANANA**	

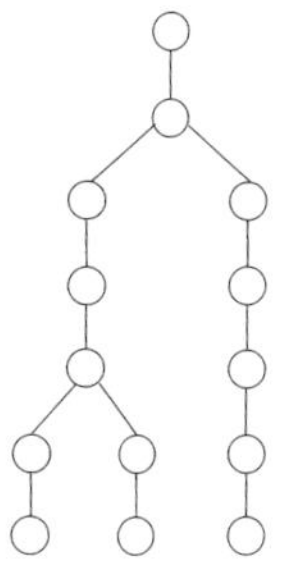

Fig. 8.8 Left: a simple taxonomy of three species in our kitchens. Right: the resulting taxonomy tree.

8.4   Classifications of species

Taxonomy is the science of ordering the biological species in a hierarchical structure where large groups contain one or more groups of smaller size. The method in its modern form was introduced into the natural sciences by Carolus Linnaeus[57] in order to classify the different living species on Earth. Different species are clustered together in genera according to similar aspect or functionality. Different genera are then clustered again into families and so on. The resulting structure is a dendrogram or a tree whose root is a group containing all living organisms. The first branching divides all the organisms into the 'kingdoms' of Mushrooms, Plants, and Animals; the last division (the leaves of the tree) corresponds to 'species'. One example of this classification tree is shown in Fig. 8.8 for three common plants. Another example could be obtained by considering the olive tree. In this case the

	species	*Olea europaea*
	genus	*Olea*
	family	*Oleaceae*
various groups are:	**order**	*Ligustrales (Lamiales)*
	class	*Dicotyledones (Magnoliopsidae)*
	phylum	*Anthophytae (Magnoliophytae)*
	kingdom	*Plantae.*

It is clear that these classifications are somehow arbitrary. The presently accepted list of levels (Greuter *et al.*, 2000) (valid only for plants!) from the

---

[57] Carl Linnaeus, also known as Carl von Linnaeus or Carolus Linnaeus was born on 23 May, 1707, at Stenbrohult, Sweden and died in 1778. His system for naming, ranking, and classifying organisms is still in use today. This classification of living things was presented in his work *Systema Naturae* (1735).

smaller to the larger are (form/variety) **species**, (section/series), **genus**, (tribe), **family, order, class, phylum, and kingdom**. The ranks in parenthesis and not in bold character are called secondary ranks and are not usually used. Intermediate levels are also possible, they are between the allowed ones and are defined by the prefix 'sub-'. Therefore a 'subclass' is the level between the order and the class. Other fine grain classifications are also used (especially for food plants) such as 'sub-species', 'variety', and 'cultivar'.

This traditional division made by Linnaeus on the basis of aspect is not the only possible one. Even keeping the aspect as the key character, different classifications can be adopted. The problem is that classifying species according to the aspect does not imply the clustering of the organisms with a similar evolution (even if, in most of the cases, two species that look similar are also genetically correlated and they share a common evolutionary history). To avoid these problems, modern taxonomy uses information from the genetic code in order to better describe also the pathogenesis of various organisms. In cladistic trees the explicit evolution of species is taken into account. Instead of grouping together plants with similar morphology, one branches the tree from a common ancestor to all the descendant species. Another possibility is that of phylogenetic trees that complement the similarities in morphology with methods taken from molecular biology.[58] Here we present only the statistical properties of Linnaean taxonomy trees for which data analysis has been extensively made. It is very likely that similar considerations might hold for any kind of taxonomic trees.

8.5   Yule process for taxonomies

The first statistical studies on this classical taxonomy trace back to the beginning of the last century. At the time it was noticed that the frequency distribution of genera $g(s)$ with s species inside is a power law $g(s) \propto s^{-\gamma}$ whose exponent is $\gamma = 1.5 \pm 0.1$ (Willis and Yule, 1922; Willis, 1922; Yule, 1925). The set considered was presented by J.C. Willis and the analysis was done by the mathematician G. Yule (we present here a modern account of the research(Aldous, 2001)). Both the original data and the plot corresponding to the model introduced by Yule are shown in Fig. 8.9.

In this model

1. The ecology starts with one genus with one species;

---

[58] The TreeBASE project at http://www.treebase.org is a repository of phylogenetic trees.

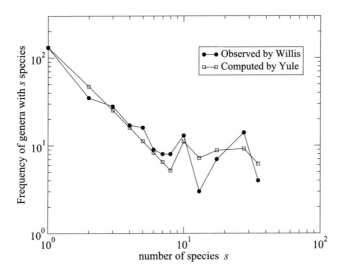

Fig. 8.9 Plot of the frequency of genera with s species for snakes. The circle are the data presented by Willis, the squares are the fitting function proposed by Yule (Aldous, 2001).

2. Starting at time 0, new species appear in the same genus. If there is one species at time t the probability to have another at $t + dt$ is given by λdt

3. From each genus a new species belonging to a new genus appear with rate μ. From this moment the genus behaves according to the previous rule.

This process (called the Yule process) is one of the basic power-law basic originators discussed in detail in Chapter 4. It can be shown that in the asymptotic limit it generates a power law for the frequency of genera with n species inside.

☢ The number of species in a genus after a time t is given simply by

$$n(t) = k_0 e^{\lambda t} \tag{8.223}$$

where k_0 for the above first hypothesis is equal to one. We want to know the distribution of species $P_k(t) = P(n(t) = k)$ at a time t. To this purpose we note that starting at time t the probability of generation between time t and $t + dt$ is given by $p_g = n(t)\lambda dt = k\lambda dt$. We can write the variation of the $P_k(t)$ in this way

$$P_k(t + dt) = P_k(t)(1 - k\lambda dt) + P_{k-1}(t)((k - 1)\lambda dt). \tag{8.224}$$

In other words we consider only two contributions: the first in which we had already k species and there is no growth (given by the term $(1 - p_g)$); the second in which we have $k - 1$ species and growth happens. In the limit $dt \to 0$ these are the important terms and we can write (by putting the first term on the right-hand side of eqn 8.224 on the left-hand side, divide both sides for dt, and taking the limit $dt \to 0$)

$$\frac{\partial P_k(t)}{dt} = -k\lambda P_k(t) + P_{k-1}(t)(k - 1)\lambda. \tag{8.225}$$

with the initial condition $P_{k_0}(0) = 1$, this has solution

$$P_k(t) = \binom{k - 1}{k - k_0} e^{\lambda k_0 t}(1 - e^{-\lambda t})^{k - k_0} \tag{8.226}$$

that for $k_0 = 1$ gives the geometric distribution with mean $e^{\lambda t}$

$$P_k(t) = e^{-\lambda t}(1 - e^{-\lambda t})^{k-1}. \tag{8.227}$$

On the other hand, the number of genera grows exponentially at rate μ. Therefore the time since the creation of a certain genus is an exponential with rate μ. Combining these two effects, we obtain that the distribution of species n in a random genus is given by

$$P_k^{all}(t) = \int_0^\infty \mu e^{-\mu t} e^{-\lambda t}(1 - e^{-\lambda t})^{k-1} dt. \tag{8.228}$$

By setting $\rho = \lambda/\mu$ we obtain a known integral (the beta integral), giving rise to the Γ functions,[59] so that we can write

$$P_k^{all}(t) = \frac{\Gamma(1 + \rho^{-1})}{\rho} \frac{\Gamma(k)}{\Gamma(k + 1 + \rho^{-1})}. \tag{8.229}$$

In the asymptotic limit we find $P(k) = P_k^{all}(t \to \infty) \propto k^{-(1+\frac{\mu}{\lambda})}$.

This model reproduces quite well the properties of the data collected. In fact, the whole set of data displays a scale-free distribution for the frequency $P(n)$ to find a genus containing a number n of species. The exponent of this distribution is around -1.5 (Willis, 1922). After that seminal study, other analyses have been done, all of them confirm to various extents the fact that the taxonomic structure has a scale-free shape. This behaviour has lately also been reported for the number $f(g)$ of families with a number of genus g and the number $o(f)$ of orders with f families, etc. (Burlando, 1990; Burlando, 1993).

---

[59] Gamma functions $\Gamma(x)$ are the extension of the factorial function. When x is a natural number n we recover the usual definition $\Gamma(k) \equiv (k - 1)! \equiv (n - 1) \cdot (n - 2)...2 \cdot 1$. When instead x is a real positive number $\Gamma(x) \equiv \int_0^\infty t^{x-1} e^{-t} dt$.

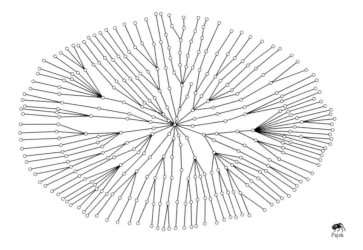

Fig. 8.10 The taxonomy tree of a small ecosystem in Amazonia.

In all the above cases the distribution of subtaxa has a fat tail and it is possible to fit them with a power-law with an exponent lying between 1.8 and 2.3. This is particular interesting since it reveals that taxonomic trees are self-similar objects.

This scale-invariance holds in different experimental situations. Traditionally, this analysis has been made on the data obtained from taxonomy books, that is by considering all the various species present on Earth, regardless of their living place. For example, the above studies are based on very large collections of species (either English flora or even more general collections). Interestingly, the same scale-invariance holds also in small ecosystems of few species irrespective of the temporal or climatic properties of the environment considered (Caretta Cartozo *et al.*, 2006). In this case one also recovers fractal properties of the structure as in the case shown in Fig. 8.10. Those results have been tested over 22 different flora from around the world with different geographical and climatic features where the number of species ranges from about 100 for specific ecosystems to about 40,000 species for the flora of entire countries. Four of the largest available food webs have been analysed (the last four in Table 8.1). In all the above cases the frequency distribution $P(n)$ for having a taxon with n subtaxa is a scale-free relation of the type $P(n) \propto n^{-\gamma}$ where the exponent γ belongs to the range 2.3 - 3.0 depending on the total number of species. The same results hold for the four food webs analysed. The same behaviour holds for temporal evolution of the same ecosystem (see Fig. 8.11). Using the data for the flora inside the Coliseum in Rome as reported from six different data collections ranging

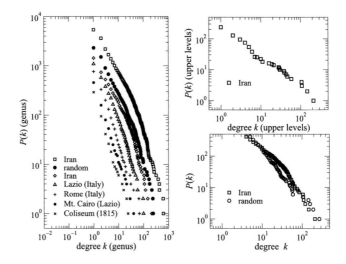

Fig. 8.11 Various data sets for both present and past flora.

from 1643 to 2001 (Caneva *et al.*, 2002), the scale-invariance is confirmed. Some of the above data are presented in Fig. 8.11.

9. Technological networks: Internet and WWW

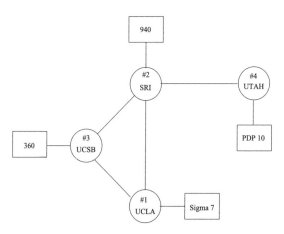

Fig. 9.1 The four nodes on the 'ARPA NETWORK', in 1969. These were University of California Los Angeles (UCLA), University of California Santa Barbara (UCSB), University of Utah, and the Stanford Research Institute (SRI). From the original picture published on http://www.cybergeography.org.

There is a saying among men, put forth long ago, that you cannot judge a mortal's life and know whether it is good or bad until he dies. Hopefully, we can pass a judgement on human activities well before their end. Without any doubt, the creation of the Internet has resulted in an entirely pacific and complete transformation of the present world. In the present society we have a simple and universal access to great amounts of information and services. Improved communication between individuals also affects everyday life changing the habits of people and offering real-time access to whatever information one might need by surfing the Web. Actually, one of the measures of the success of the World Wide Web (WWW) is that people consider

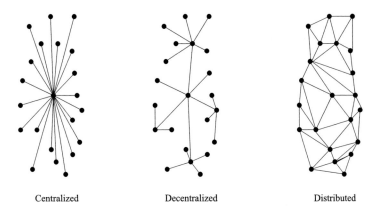

| Centralized | Decentralized | Distributed |

Fig. 9.2 The various possibilities for a communications network (Baran, 1964). The last one is the structure programmed for the Internet.

WWW and the Internet as synonymous. This is technically wrong. As we see in the following, the Internet is the system composed of 'computers' connected by cables or some other form of physical connection. Over this physical network it is possible to exchange e-mails, transfer files, or visualize some of them. One of the services defined over the layer of the Internet is that of the WWW.

As the Internet project started, the idea was to establish a series of communication lines that could resist to a military strike. Later, a similar structure was adopted to allow scientific communication between different institutions (see Fig. 9.1). In a seminal paper of the 1960s (Baran, 1964) the properties that such network must have were already envisaged. The basic idea is to have what is called a distributed system,[60] as depicted in the third image of Fig. 9.2. Following these preliminary studies the structure of Internet started to be developed in 1969. To ensure communication during a military strike it was necessary to define a special way to deliver information. The solution was to design a dynamical routing for the messages in order to redirect the traffic according to the situation of the network. The first prototype of the system of communication (called ARPANET) was realized by the Defense Advanced Research Project agency. This system of communication was based on the first version of the present protocol (set of fixed rules) TCP/IP used to exchange files and to connect to other computers. By 1983 562 hosts were registered in ARPANET. Three years later, the US National Science Foundation started the project 'Supercomputer Center'.

---

[60] Note that in this chapter we provide evidence that the present shape is actually more similar to the second one (the decentralized one).

This project, connecting the main scientific organizations in the country in order to allow exchange of data and ideas, resulted in an effective larger network. In the 1990s, thanks to the explosion of the WWW the growth of the Internet has become exponential. At the moment (2006) the number of computers connected to each other is still growing.

The growth of web pages has been even larger than that. At beginning of 1998 (Lawrence and Giles, 1998) one estimate put it at more than 320 million pages (Lawrence and Giles, 1998). At the end of 1999 the number of vertices was around one billion (Lawrence and Giles, 1999). Presently this number is probably much larger. Note that the concept of 'total number of pages' in the WWW is ill-posed. Many pages do not exist in the traditional sense. In order to be counted as a vertex of the network a page should be stored somewhere and be accessible by typing a specific address on the browser. However, many pages are created upon user request and disappear after they are consulted. Note also that a rapid decay of pages is associated to this exponential growth (Bar-Yossef *et al.*, 2004). All we can say is that on average 'stable' traditional pages are still growing year after year. One rough estimate is provided by the search engine Google. At the moment of writing, they claim to have a database of more than 8 billion different pages.

As regards e-mails hundreds of millions (if not billions) of people use this service. This is another service provided by the Internet layer. Anyone connected to the Internet can compose a document with text, images, music, or whatever and send it to another user if the latter has an e-mail address. This means that the document is transferred (in some complicated way) to another computer indicated by the e-mail address of the recipient. Once the message is delivered to this computer it is stored in a special directory that is the user's mailbox. The user can access the file after some sort of agreement (usually commercial) with the owner of the computer destination (provider of the service). The order and size of this network are smaller but still comparable to those of the WWW.

We present in the following sections some experimental analysis for all these systems. The graph of the Internet has a particularly intuitive meaning. Computers are the vertices of this system; cables or any other means of transmission represent the various edges. The graph is undirected, since messages can travel back and forth on the same edge. It is important to note that the edges of the graph do not share similar properties. They represent the Internet 'cables' and therefore they can differ in capacity, speed, and cost of use. In principle the best Internet picture would be given by a weighted undirected graph. Since traffic statistics are very difficult to obtain, most of the studies so far consider only the Internet topology. As a result, almost any model of the Internet represent it as an undirected graph where no weights are attached to edges or vertices. Another problem is that this graph is (even if loosely) embedded in a truly three-dimensional system. In an ordinary graph the distance between vertices is given by the number of edges separating them. Here instead vertices occupy geographical positions. Edges between computers in the same building are far more probable than edges between a personal computer (PC) in Europe and another in Australia. It is then possible to detect some effects of geographical location on the features of the network (Morita, 2006). Amongst the different networks presented in this book, the Internet is the only one where metric distances are likely to affect the probabilities of drawing or not an edge. The WWW instead is a directed graph since a hyperlink transfers from a specific page to another but not vice versa. E-mails are also better described by a directed graph. Spam mail is an example of messages we often receive without answering back. We present in the following a brief description of the physical structure of the system and the various models that have been proposed. For the latter, the overwhelming majority of studies makes the approximation of describing the Internet as a non-metrical unweighted and undirected graph. Further reading about these communication networks is presented in specialized texts (e.g. Pastor-Satorras and Vespignani 2004).

> In this chapter we present the latest studies on the Internet network and on the World Wide Web which is a 'software' network defined over the Internet layer. The peculiarity of these systems is their gigantic size, of the order of billions of edges and vertices. This size makes it very difficult to measure their topological properties. Even worse, technological networks are also highly dynamic, such that the actual structure of both Internet and WWW changes on timescale of hours.

9.1   The Internet protocols

Once we have more than one computer, it is theoretically possible to communicate, provided the computers 'speak' a common language. For most common functions such as the exchange of e-mail, file transfer, and the remote login, there is a *suite* of protocols that we describe briefly. The core of these controls are the Internet Protocol (IP) and the Transmission Control Protocol (TCP). They define the address of the various computers and establish the way in which information is transmitted. For example, both for practical (the same computer can make different operation at the same

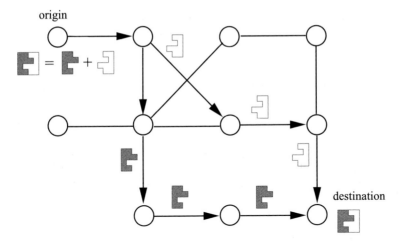

origin

destination

Fig. 9.3 The delivery process. A document is divided and sent away. Step by step every router in the path decides autonomously the next router towards the destination. Note that at the same time other packets travel through the same servers to different destinations.

time) and security reasons (only at destination we find the whole document) the documents are divided in packets before being transmitted as shown in Fig. 9.3. In general, the IP has:

- a set of rules assigning a unique address to various hosts (see Section 9.1.1);
- a set of rules to fragment any document into little portions. In these 'packets' is specified the sending and addressing node, the length of the packet, its position in the original data (non-fragmented), and various other pieces of information.

To avoid some fragments travelling endlessly around the net, every packet also has a specified 'time to live' to reach its destination. When the time to live is exceeded, the packet is discarded. The whole process is like sending the pieces of a puzzle through postal mail. The whole puzzle (the document) is fragmented and each fragment is put in an envelope with sender and destination addresses (Internet 'envelopes' also provide other information in order to complete the puzzle faster!). The reliability of such a procedure is ensured by the TCP. This piece of software is in charge of suitably chopping the data into little fragments. TCP also indicates the set of fixed signals that two hosts must exchange to exchange information. This transfer is made in such a way that every host is both transmitting and receiving packets relative to different documents. A list and a detailed explanation of these rules is beyond the scope of this book. Here we only provide some information on this protocol of communication. While other protocols exist, TCP remains one of the most important since it defines a series of services like the HyperText Transfer Protocol (HTTP), the File Transfer Protocol (FTP), and Telnet. These services ensure the functioning of the most popular operations on the net like navigation of the World Wide Web and the exchange of files.

9.1.1 Labelling the vertices: IP address

To allow any operation in the Internet, the various computers must be identified uniquely. This means that one node must have an address allowing it to be distinguished from all the others. The computer on which I am writing at this moment has the Internet Protocol (IP) address, 151.100.123.37. In general, IP addresses have this structure of four decimal numbers separated by periods. The choice for the numbers is limited: they must be between 0 and 255. This immediately gives the total number of possible different addresses. They are $256 \times 256 \times 256 \times 256$ or in a compact notation 256^4. Since 256 is equal to 2^8 we have that the total number of addresses can be also expressed as 2^{32} (the address is technically indicated as a 32-bit digit), namely above number corresponds to more than 4 billion $(4, 294, 967, 296)$ addresses available (in principle).

   The situation is slightly more complicated, and the above figure is loosely related to the total number of PCs forming the net. Firstly, different computers can be connected together in an office where they receive a 'private address'. This means that they form a local network connected to the external net through a special computer called a 'firewall'. In this case only the firewall has a valid Internet address. Thus, one address takes then into account a whole subnet of computers rather than a single PC. Secondly, the address could correspond to a printer or to another device. For this reason, scientists usually refer to *hosts* attached to a certain address. Note that the structure of the IP address scheme is hierarchical and not on a 'first come, first served' basis. This means that the oldest host connected to the net does not have the IP address 0.0.0.0. Rather the address (expressed in its binary form) reports some information about the type of 'local' network where the host is present. For all these reasons, the size of the Internet can only be roughly estimated from the number of used IP addresses (which are currently running out).

The binary form[61] of the IP address is important because it determines which of the five possible classes of network the IP address belongs to. In any case one part of the address identifies the network and the rest identifies the node, or host. The various classes differ for the proportion of space indicating who is the network and who is the host.

- **Class A** networks have a binary address starting with 0. The decimal number of the first group goes from 1 to 126. The first 8 bits (the first group) identify the network and the remaining 24 bits (the other three groups) indicate the host in the network. The address in decimal numbers is of the form *N.H.H.H* where *N* indicates the network and *H* the host. This is a class used by a few large organizations since it allows a maximum of $16,777,214$ hosts ($2^{24} - 2$) per network.
- **Class B** networks have a binary address starting with 10. The decimal number of the first group goes from 128 to 191 (127 is left out and reserved). The first 16 bits (the first two groups) give the network and the other 16 bits (the last two groups) give the host within the network. The address now has the form *N.N.H.H* and this is the typical class of medium-sized organizations such as my university (indeed my PC has address 151.100.123.37). The network belonging to the department of physics in my building has network identification 151.100 and the host has identification 123.37. The maximum number of hosts per network is now $65,543$ ($2^{16} - 2$) a pretty large number even for a department of physics.
- **Class C** networks have a binary address starting with 111. The decimal number goes from 192 to 223. The first 24 bits (the first three groups) identify the network and the remaining 8 bits indicate the host within the network. The form of the address is *N.N.N.H*. Mainly used by relatively small organizations, it allows 245 ($2^8 - 2$) hosts per network.
- **Class D** networks have a binary address starting with 1110, therefore the decimal number can be anywhere from 224 to 239. Class D networks are used to support multicasting.
- **Class E** networks have a binary addresses starting with 1111, therefore the decimal number can be anywhere from 240 to 255. Class E networks

---

[61] A number expressed in binary form is a series of digits 0 and 1 giving the number of powers of 2 in which the number can be decomposed. The number 110 in binary form does not mean 1 hundred (10^2) plus 1 tens (10^1) plus 0 units (10^0), but rather 1 group of four (2^2) plus 1 couple (2^1) plus 0 units (2^0), that is 6. Internet addresses are made by four groups of 8-bit numbers. That is, we have four times a number like *xxxxxxxx* where *x* can be either 0 or 1. The largest possible 8-digit number in binary is 11111111; in the decimal representation this corresponds to 255.

are used for experimentation and have not been documented or utilized in a standard way.

We also know that computers can be indicated by more conventional names. For example the computer on which I am writing now is also known as falerno.phys2.uniroma1.it. A domain name server (DNS) is a computer that relates these conventional names with the corresponding IP addresses. The same mechanism works for web pages. The reason for the double standard is because we are more inclined to remember words and concepts than a series of numbers. In this way if we want to access the Oxford University Press page, we type on our browser *http://www.oup.com*, where the *http* declares the type of services we want (HyperText Transfer Protocol), *www* reminds once again that we are looking for web pages, *oup* is the acronym of the company (*com*). This is somewhat easy to remember than the corresponding IP number 12.107.205.41. Furthermore, if the server where the information is stored needs to be changed or repaired, the web page could still be accessible using another server and updating the DNS. Since computers do not mind using numbers instead of words, they can work with both. If we type into the browser the above numerical address (i.e. http://12.107.205.41) we reach the same page (without checking a DNS).

9.1.2   Routing in the Internet

Once you have addresses for the hosts, you must specify how they connect each other. The protocols for routing are a set of rules that bring the packets to their destination step by step. They are simply a list of two entries: destination addresses and next step path. An example is

151.100.123.37   151.100.123.1

that is to say when the router receives a packet with destination on the first column it 'knows' it has to deliver it to the address on the second column. Things can change and the list present in any router (a computer whose specific action is to transmit packets) is dynamically adjusted at regular intervals (less than minute).

Any router communicates with its immediate neighbours; eventually this information propagates through the net. The mechanism is similar to what is done when looking for a place in an unknown city. If we ask information of the first person we meet very likely we obtain a vague indication of the direction. By successive steps, we gain more and more precise indications and we reach our destination. In this process something can go wrong. If packets get lost they die out along the road, the document is not delivered and another protocol, IMP (Internet control Message Protocol), reports the error. This delivering one hop at time allows an exploration of the Internet through the traceroute program as shown in the next section.

9.2   The geography of the Internet

Drawing a map is the best way to find our way in a new environment. Usually the maps are easy to draw and read. They report the distance between places so that once we have identified the origin and destination we can find immediately the distance and the best path between them. In the case of the Internet, every single step is an edge between two routers. It does not matter if one is located next door and another is 10 kilometres away. Therefore, distance becomes a graph distance, that is it is the number of edges (hops) between two sites. Actually, a heritage of the physical distance between routers is still present in the Internet. Cables are very expensive, this means that we can afford to connect one PC to another one nearby, but very seldom do we create new edges (physical lines) of hundreds or thousands of kilometres. Anyway, extracting the metrical information from the net is rather tricky (Lakhina *et al.*, 2003; Huffaker *et al.*, 2002) and for the moment we can think of the Internet as a true graph, where metric distance does not play a relevant role.

One software tool designed to explore the net is *traceroute*. This software sends a series of packets with a specified time of life to a specific destination. Every router receiving the packet is instructed to decrease the time of life and to send back a message of acknowledgement. The collection of these messages of acknowledgement allows us to reconstruct the path followed by the packets. This software tool is available from the command shell of Microsoft Windows, Apple operating system, and the various dialect of Linux. A typical example of traceroute from my PC in Rome to another PC in Paris is the following.

```
[gcalda@falerno gcalda]> traceroute tournesol.lps.ens.fr
traceroute to tournesol.lps.ens.fr (129.199.120.1), 30 hops max,
38 byte packets
 1  151.100.123.1 (151.100.123.1)  1.858 ms  1.456 ms  1.739 ms
 2  151.100.222.1 (151.100.222.1)  6.607 ms  6.101 ms  5.863 ms
 3  151.100.254.250 (151.100.254.250)  4.984 ms  4.852 ms
    5.115 ms
 4  rtg-unirm1-2.rm.garr.net (193.206.131.237)  6.985 ms
    5.970 ms  5.612 ms
 5  rt-rm1-rt-mi2.mi2.garr.net (193.206.134.229)  14.230 ms
    14.958 ms  14.234 ms
```

```
 6   rt-mi2-rt-mi1.mi1.garr.net (193.206.134.17)  22.845 ms
     233.706 ms   20.600 ms
 7   garr.it1.it.geant.net (62.40.103.189)  14.602 ms   14.467 ms
     14.607 ms
 8   it.ch1.ch.geant.net (62.40.96.33)  27.982 ms   27.975 ms
     28.487 ms
 9   ch.fr1.fr.geant.net (62.40.96.30)  36.481 ms   36.977 ms
     35.986 ms
10   renater-10G-gw.fr1.fr.geant.net (62.40.103.162)  36.480 ms
     36.974 ms   36.485 ms
11   nri-c-g4-0-0.cssi.renater.fr (193.51.179.38)  37.982 ms
     43.969 ms   37.485 ms MPLS Label=334 CoS=0 TTL=1 S=1
12   jussieu-pos4-0.cssi.renater.fr (193.51.180.157)  65.476 ms
     38.471 ms   35.985 ms
13   193.50.20.73 (193.50.20.73)  36.484 ms   36.975 ms   38.485 ms
14   cr-odeon.rap.prd.fr (195.221.127.109)  38.482 ms   37.473 ms
     36.985 ms
15   ens-ulm.rap.prd.fr (195.221.127.62)  38.482 ms   49.970 ms
     46.984 ms
16   merlin-gw.ens.fr (129.199.1.16)  39.983 ms   38.474 ms
     37.985 ms
17   tournesol.lps.ens.fr (129.199.120.1)  53.981 ms   38.473 ms
     38.983 ms
```

The first field is the number of hops from the origin node, the second field is the name of the host crossed. The latter is provided both with 'names' (when available) and with IP addresses. The last fields report the round trip time in milliseconds (ms). This latter quantity is a measure of the time needed to reach destination.

The Internet Mapping Project sends parallel probes towards about 10^6 possible addresses, and reconstructs maps of the kind shown in Fig. 9.4. A similar analysis is in principle possible from any computer, even if we suggest that readers do not create their own map of the net.[62] This procedure gives a biased map since it detects only the direct path from one host to the others and gives little or no information on the connections between the various destination hosts. Therefore, by changing the origin we obtain

---

[62] Note that according to the Linux manual 'traceroute is intended for use in network testing, measurement and management. It should be used primarily for manual fault isolation. Because of the load it could impose on the network, it is unwise to use traceroute during normal operations or from automated scripts'.

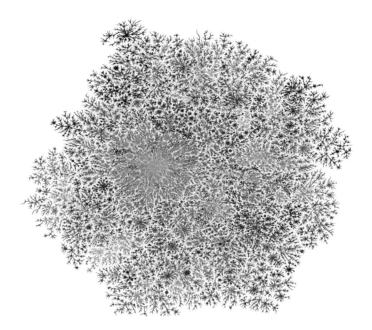

Fig. 9.4 A map of the Internet realized by the Internet Mapping Project (http://research.lumeta.com/ches/map/). This map has different grey levels according to the distance from the source of traceroute (courtesy of Lumeta).

different maps. An interesting scientific problem is then to relate the properties of traceroute maps with the properties of the whole network. Some studies try to determine the minimum set of sources in order to obtain a fair representation of the statistical properties of the whole (Petermann and De Los Rios, 2004). Others focus on the change in statistical properties. It has been shown analytically (Clauset and Moore, 2005) that, for random graphs with mean degree (Poisson distributed) c, a traceroute sampling gives an observed degree distribution $P(k) \propto 1/k$ for $k < c$. For graphs with power-law degree distributions $(P(k) \propto k^{-\gamma})$, traceroute sampling from a small number of sources can significantly underestimate the value of the exponent. Further analysis has confirmed that in general while traceroute can give a fairly correct qualitative description of the whole system, it can nevertheless introduce large deviations from the correct quantitative behaviour (Dall'Asta *et al.*, 2005).

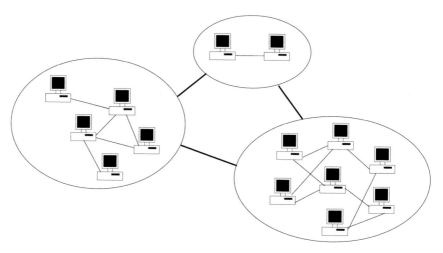

Fig. 9.5 A sketch of the structure of three small autonomous systems

9.3   The autonomous systems

The autonomous systems (AS) are the first elementary partition of the Internet. This partition divides the Internet into different subgraphs whose control for technical and bureaucratic reasons is given to a local administrator. They can exercise restrictions of various kinds, on traffic, bandwidth, and policies. Autonomous systems vary of course in size and function. Some of them (the *transit* ones) are in charge of national or international connectivity. On a local scale *stub* AS are those providing connections to companies, departments, and local areas. As a matter of fact, the situation is slightly more complicated, with stubs connected each other. The AS may be further divided into domains, regions, or areas that define a hierarchy within the AS. The protocols used to distribute routing information within an AS are known as IGP (Interior Gateway Protocols). The protocols used to distribute routing information between different AS are known as EGP (Exterior Gateway Protocols).

Using the technical language we can define an AS graph of the Internet as a network where AS represent the vertices and the connection between different AS represent the edges (see Fig. 9.5). This first coarse-grained description of the system allows us to handle the complexity of the systems by reducing the size of the graph. Almost all the statistical analysis of Internet

	γ	β	l	C
AS 1999	2.2 ± 0.1	1.0	3.7 ± 0.1	0.24 ± 0.3
AS 2004 skitter	2.25	1.35	3.12	0.46
AS 2004 BGP	2.16	1.17	3.69	0.29

Table 9.1 A list of the main topological quantities for three different AS maps, as measured by different teams. The exponents γ and β are those of the power-law distributions of degree and betweenness respectively, l is the average distance in the network, and C is the clustering coefficient.

presented here is based on AS graphs. Different repositories for AS maps are available on the net.[63]

9.4   The scale-invariance in the Internet

One of the first studies on these statistical distribution was made on a subset of Internet routers and autonomous systems (Faloutsos, Faloutsos, and Faloutsos, 1999). One of the main findings is that the degree probability distribution is characterized by heavy tails that can be fitted by a power law. After that, several other measurements (Bu and Towsley, 2002) and analyses (Caldarelli, Marchetti, and Pietronero, 2000; Vázquez, Pastor-Satorras, and Vespignani, 2002) have been made on the Internet graph, always confirming the presence of scale-free behaviour. A sketch of the main measurements is reported in Table 9.1. One of the typical comment, in the case of the Internet is that this scale-invariance is probably related to the self-organization of the object. True or not, it is interesting to note that this is not what the founders of Internet planned for the development of their creation. Almost surprisingly, scale-free networks were already considered under the name of 'decentralized networks' (see Fig. 9.2) and discarded as suitable models for Internet growth. Others believe that scale-free behaviour might be an artefact, since the traceroute procedure for collecting data introduces a systematic bias. Actually, scale-invariance is also confirmed by many studies on Internet maps based on the autonomous systems structure (Gao, 2001; Di Battista, Patrignani, and Pizzonia, 2002).

AS graphs are formed by using the data of the Border Gateway Protocol (BGP), specifically suited to connect different ASs. In this data we find the various connections at the host level. By grouping together the hosts

---

[63] http://www.cosinproject.org/extra/data/internet and http://www.caida.org/tools/measurement/skitter/as_adjacencies.xml

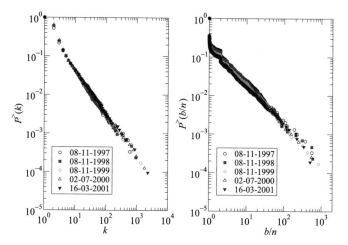

Fig. 9.6 The plot of the frequency distribution for the degree (left) and for the normalized betweenness (right) for the autonomous systems maps. Courtesy of A. Vazquez (Vázquez, Pastor-Satorras, and Vespignani, 2002).

under the same domain one obtains the AS-level representation. Results from recent studies (Mahadevan *et al.*, 2005) from CAIDA suggest that topological properties of the graph obtained from traceroute and BGP are substantially similar. Differences exist when the source of data is the WHOIS data set. WHOIS is a manually maintained collection of databases. Some entries are relatives to connections between AS and those are used to build the map. This supports the claim that the Internet is self-organized in a stationary state characterized by scale-free fluctuations (Vázquez, Pastor-Satorras, and Vespignani, 2002).

The first topological quantity is the integrated frequency distribution $P^>(k)$ for the degree k, shown on the left of Fig. 9.6. We have a power law

$$P^>(k) \propto k^{1-\gamma}, \tag{9.230}$$

where the exponent γ can be found as 2.2 ± 0.1 from various experiments.

A similar value is obtained for the plot of the integrated frequency distribution $P^>(b_n)$ for the normalized betweenness $b_n = b/n$ where n is the number of vertices of the graph. Also in this case (Vázquez, Pastor-Satorras, and Vespignani, 2002) we find a power law of the kind

$$P^>(b_n) \propto b_n^{1-\delta}. \tag{9.231}$$

The value of δ is very similar to that of γ yielding $\delta = 2.1 \pm 0.1$. As regards the relation between the betweenness b of a vertex and its degree k one finds again a power law, whose exponent can be computed from the

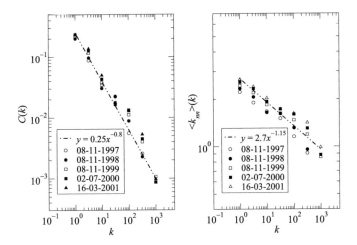

Fig. 9.7 The plot of the clustering coefficient of vertices whose degree is k and the average value of the neighbours degree (right) (Vázquez, Pastor-Satorras, and Vespignani, 2002).

previous exponents. This is one simple example of scaling relations between exponents in the Internet. If we assume that the betweenness is related to the degree by means of an expression like $b_n(k) \propto k^\beta$, by substituting $b(k)$ with k^β in eqn 9.231 and comparing the expression with that of eqn 9.230, we have

$$\beta = \frac{1 - \gamma}{1 - \delta}. \tag{9.232}$$

Since the two exponents γ and δ have a similar value we must find $\beta \simeq 1$. This is confirmed by experimental analysis (Mahadevan *et al.*, 2005) whose findings give $\beta = 1.15\text{-}1.37$.

As regards hierarchy and assortativity measurements, a comparison over different databases for different years also gives the same results. In Fig. 9.7 we report the data sets analysis as well as the power-law fitting functions proposed.

9.5 The World Wide Web

One of the most successful services working on the Internet network is that of the World Wide Web (WWW). This method of disseminating information was invented at CERN (Conseil Européen pour la Recherche Nucléaire) by physicist Tim Berners-Lee in 1989. The idea was to allow scientists in different places to share information automatically. The core of the idea is to allow many cross-references between texts (hyperlinks) and create a web of knowledge that could be browsed by readers. In the case of the WWW the directed graph is composed by the documents or web pages (vertices) connected by hyperlinks (directed edges). The directed structure reflects the fact that a person may put many hyperlinks in his/her page, but seldom or never (as in the case of the pages of newspapers or flight companies) do these pages draw a reciprocal hyperlink to the users. A web page is a text document written in a special language called HyperText Markup Language (HTML). In the simplest form this file is nothing more than text that will appear on the screen of computers when visualized by specific programs called browsers. Every document can be accessed if it is placed in a specific place on special computers called web servers. To access the document we need to know its address. Web addresses are different from Internet ones; we have already seen that specific computers (domain name servers) map one type of address to the other. In this way when we type http://www.oup.com into a browser, this software 'knows' that it has to ask to its DNS what is the Internet address related to that web address (presently at 12.107.205.41). After that the browser starts a protocol of communication between computers called HyperText Transfer Protocol (http), which contacts the remote computer, accesses the contents of the remote hard disk, looks into the appropriate directory, and displays the relevant page.

9.6   Searching the web

9.6.1   HITS algorithm, hubs and authorities

The WWW presents distinctive features as compared to other real networks. As noticed above, it is a directed graph, where the vertices may be very different from each other. This division is based on contents of the web pages and as a first approximation they can be divided in two different classes. This division reflects the intrinsic nature of HTML documents. In the first case one page can be created in order to provide specific information on some topic. This can be the list of players of our favourite team, the address of a shop, a picture of our city. On the other hand there are other pages (essentially a list of hyperlinks) whose only aim is to help users find specific information (e.g. the Yahoo search engine at http://www.yahoo.com). In principle, we can distinguish between these two classes by means of content analysis. For example, a page with texts or figures and few hyperlinks is probably devoted to disseminating information. A page whose only contents

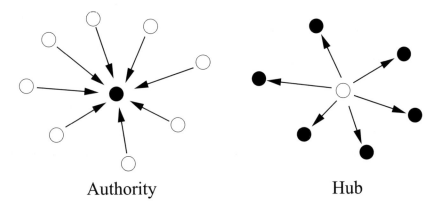

Fig. 9.8 The nature of the edges in an authority page and a hub page.

are hyperlinks is probably devoted to presentation of other pages. In practice, it is possible to use the topology of the graph to discover the web page nature. The most successful pages with specific information inside (teams, addresses, pictures) have many hyperlinks pointing to them and few going out. On the contrary, listing pages have many outgoing hyperlinks (only if very successful also many ingoing hyperlinks). In the first case, we call the page an **authority** ; in the second case, we call the page a **hub**. Almost all pages have a double nature because the best hubs are for this reason also authorities in their field. The relation between these two kinds of pages is shown in Fig. 9.8.

The way to compute the 'hubness' and 'authoritativeness' of a page is self-consistent and is at the basis of the algorithm for Hypertext Induced Topic Search (HITS) (Kleinberg, 1999a). The idea is that a site pointed by a very good hub will be more authoritative than one pointed by a very bad hub. As already explained in Section 2.6.3, we can measure the authoritativeness x_i and hubness y_i of every vertex i in a given graph. We repeat here the derivation of the measure for such quantities. Let us start by considering a subgraph of the web from which we remove all the edges internal to the same domain (as those to the webmaster). Any edge from page j to i increases the authority of page i. Every vertex i (a page) has a non-negative authority weight x_i and a non-negative hub weight y_i that we want to compute. Those two quantities are related by the formulas

$$x_i = \sum_{j \to i} y_j$$

$$y_i = \sum_{i \to j} x_j \qquad (9.233)$$

where $j \rightarrow i$ runs on all the pages j pointing to i and $i \rightarrow j$ runs on all the pages j pointed by i.

 In the matrix formalism, this can be written as

$$\mathbf{x} = \mathbf{A}^{\mathbf{T}}\mathbf{y}$$

$$\mathbf{y} = \mathbf{A}\mathbf{x} \qquad (9.234)$$

where A is the adjacency matrix and A^T is its transposition. By simple substitution of the second equation in the first, we obtain the relations

$$\mathbf{x} = (\mathbf{A}^{\mathbf{T}}\mathbf{A})\mathbf{x}$$

$$\mathbf{y} = (\mathbf{A}\mathbf{A}^{\mathbf{T}})\mathbf{y} \qquad (9.235)$$

stating that the set of values x_i corresponds to the eigenvectors of the matrix $\mathbf{A}^{\mathbf{T}}\mathbf{A}$ (the authority matrix) while the values y_i correspond to the eigenvectors of matrix $\mathbf{A}\mathbf{A}^{\mathbf{T}}$ (the hub matrix) where A is the adjacency matrix of the graph.

Amongst the different ways to solve this system, the most used is by recursion. That is, we start from fixed initial values x_i^0, y_i^0 that we put on the right-hand side of eqns 9.233; from that one obtains the new values x_i^1, y_i^1. Those values are then put again on the right-hand side of the same equations obtaining the values x_i^2, y_i^2. After a certain number t of iterations the values x_i^t, y_i^t differ only for a tiny amount from x_i^{t-1}, y_i^{t-1}. This amount can be further reduced by making some more iterations.

In the case described it can be demonstrated that under certain hypotheses this procedure converges to the solution required.[64]

The first simple requirement for convergence is to normalize the vectors $\mathbf{x^t}, \mathbf{y^t}$ at any iteration (i.e. to impose that $\sum_{i=1,N} x_i^2 = 1$). A less simple requirement is the irreducibility of the hub and authority matrices. We describe this problem in detail in the next section. Without entering into details we mention that some modifications of the HITS algorithm have been presented in order to overcome these problems (Farahat *et al.*, 2006; Miller *et al.*, 2001).

9.6.2   The PageRank

The concept of PageRank (Page *et al.*, 1999) is at the heart of the search engine Google (http://www.google.com). Through this method, the search

---

[64] The matrices $A^T A$ and AA^T are symmetric, positive semidefinite, and non-negative. Henceforth their distinct eigenvalues are real and nonnegative so that we can sort them $(\lambda_1 > \lambda_2 > ... > \lambda_n \geq 0)$ (Demonstration can be found in basic texts of linear algebra (e.g. Golub and Van Loan 1989)).

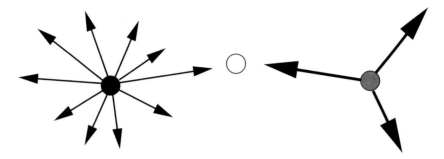

Fig. 9.9 The idea of the PageRank. Edges from good sites (grey vertex) with low out-degree contribute more than edges from bad sites (black vertex) with large out-degree.

engine determines the page's relevance or importance. For an overview of this method (as well as of HITS) there are some extensive reviews (e.g. Langville and Meyer 2005; Rogers 2005; Langville and Meyer 2006). The PageRank can be viewed as a vote for a document given by the other pages on the Web. Every hyperlink received by a web page counts as a vote, therefore the larger the number of hyperlinks the larger the number of votes received. Differently from a democratic process, not all the votes count the same. Rather the votes of other pages are weighted by their own PageRank. This means that the larger is the PageRank of the voting page, the larger its effect on the page destination, as shown in Fig. 9.9. If one page has many outgoing hyperlinks we can assume that the authority is divided among the various destinations. This corresponds to saying that PageRank is normalized by the out-degree of the pages.

By using all these ideas we can provide a first preliminary definition for the PageRank r_i of a page

$$r_i = \sum_{j \to i} \frac{r_j}{k_j}. \tag{9.236}$$

This equation can be conveniently rewritten by use of linear algebra as

$$\mathbf{r} = \mathbf{r}\mathbf{N}. \tag{9.237}$$

Here $\mathbf{N}$ is the normalized matrix whose elements n_{ij} are 0 if there is no edge between i and j and $1/k_i$ otherwise (see Section 2.6.2.1). Alternatively, we can write $\mathbf{N} = \mathbf{A}\mathbf{K}^{-1}$ where $\mathbf{A}$ is the adjacency matrix, and $\mathbf{K}$ is a diagonal matrix whose non-zero entries k_{ii} are the degrees of vertices i.

While this is very simple in theory, in practice it is very difficult to implement. Consider a real portion of the World Wide Web: almost certainly there will be pages without outgoing edges (sometime referred as *dangling*

Fig. 9.10 A simple example of a trapping state.

nodes). This results in rows whose entries are all zero. These dangling nodes give no contribution to rank determination and can be deleted. When dangling nodes are discarded, all the rows of matrix N have non-zero entries.[65] The matrix can now be viewed as a stochastic matrix giving the evolution of a random walk on the graph. A random walk is a walk where every step is chosen randomly amongst the various possibilities. Consider that you are on a certain vertex i, you can move along the graph to neighbouring vertices with a probability given by $1/k_i$ for each of the k_i edges. This is exactly the transformation represented by $\mathbf{N}$. For this reason the PageRank also has the meaning of the time spent on a page by a surfer who chooses a new page randomly amongst the available edges.

A practical problem is that the order of the matrix is really huge, with billions of rows and columns; any analytical solution of the eigenvalue problem is impossible. Therefore in this case we have to use the power method and solve the system by iteration. This corresponds to assigning a random rank $r(i)$ to the pages of the graph and iteratively applying the matrix N in this way

$$\mathbf{r^t} = \mathbf{r^{t-1}N}. \tag{9.238}$$

The presence of dangling nodes avoids $\mathbf{N}$ being a stochastic matrix and therefore gives problems for the existence of the limiting vector $\mathbf{r}^\infty$, which represents the numerical solution of the equation $\mathbf{r} = \mathbf{rN}$. Even worst almost certainly the subgraph represented by $\mathbf{N}$ will be reducible. A reducible stochastic matrix is one for which the underlying chain of transformations is reducible.[66] A reducible chain, is one for which there are states in which the evolution can be trapped. The simplest example is that of a page i that has only an edge to page j and this page j has only an edge to page i (see

---

[65] Actually the dangling nodes are usually considered in the matrix N. Instead of using their real out-degree (that is zero), one makes a little perturbation by substituting the zero entries with a vanishing value $1/n$ where n is the order of the matrix. The order of matrix in this case is the number of sites in the portion of the subgraph analysed (dangling nodes included).

[66] In this case the chain is called a Markov chain, since the state at a certain time of evolution depends only upon the state at the previous time step.

Fig. 9.10). Iteration on this set will not produce convergence to a limit vector r^∞, but rather an oscillation between the two states. When the matrix is *irreducible* a mathematical theorem (by Perron and Frobenius) ensures that this chain must have a unique and positive stationary vector r^∞. A way to force irreducibility numerically is to destroy the possibility of getting trapped. If you can jump out from a page to a completely random different one (even with small probability) the matrix is irreducible and you can find the eigenvectors r by iteration. This corresponds to adding to the matrix N another diagonal matrix E whose entries e_{ii} are given by $1/n$ where n is the number of vertices in the graph.[67]

Therefore the final definition of PageRank is the following:

The values of PageRank for the various pages in the graph is given by the eigenvector r related to the largest eigenvalue λ_1 of the matrix P given by

$$P = \alpha N + (1 - \alpha)E, \tag{9.239}$$

the weight α is taken as $\alpha = 0.85$ in the original paper (Page, Brin, Motwami, and Winograd, 1999).

This new matrix P does not differ considerably from the original one N, but has the advantage that (thanks to its irreducibility) its eigenvectors can be computed by a simple iteration procedure. For a complete description of the implementation of PageRank see Langville and Meyer (2005).

9.7   Statistical measures of the Web

Also in the case of the WWW we deal with scale-invariant degree distributions (Albert, Jeong, and Barabási 1999; Kumar *et al.* 1999*b*; Kumar *et al.* 2000; Broder *et al.* 2000; Laura *et al.* 2003). Since the edges are directed the distributions to be considered are both the out-degree ($P^o(k^{out})$) and the in-degree ($P^i(k^{in})$) one. For them we find respectively:

$$P^o(k^{out}) \sim (k^{out})^{-\gamma^{out}} \quad \text{and} \quad P^i(k^{in}) \sim (k^{in})^{-\gamma^{in}}, \tag{9.240}$$

with a value of $\gamma^{in} = 2.1$ in most of the cases and a value $\gamma^{out} = 2.4\text{-}2.7$.

These analyses have been made in different years and using very different data sets, thereby confirming the intrinsic scale-invariance of the Web. One of the first analysis consists of a set of $325,729$ pages and $1,469,680$ hyperlinks in the domain of the University of Notre Dame (Albert, Jeong, and

---

[67] In the more recent implementation of PageRank, those entries are actually different each other even if they have the same order of magnitude. This is done in order to introduce an ad hoc weight for the different pages.

Barabási, 1999). Immediately following (literally) another study (Huberman and Adamic, 1999) confirmed the power-law observed on two different data sets of about $259,794$ and $525,882$ pages. In a sequent analysis (Adamic and Huberman, 2000), the authors adopted a clustering approach to overcome the problem of the graph size. They used a different representation of the WWW, where every node represents a separate domain name and two nodes are connected if any of the pages in one domain is linked to any page in the other. This method often collapses into one vertex thousands of pages that are on the same domain, but surprisingly the distribution $P^i(k^{in})$ is still a power law with $\gamma_{dom}^{in} = 1.94$. At this domain level the diameter was found consistently smaller with a value $d = 3.1$. The clustering coefficient of the WWW has also been considered (Adamic, 1999). To overcome the problem of the definition of the directed clustering coefficient (see Section 1.1.2) they considered an undirected network. They analysed a data set of 50 million web pages distributed between $259,794$ sites by removing vertices with degree $k = 1$, thereby reducing the number of vertices to $153,127$. They found a $C = 0.1078$ to be compared with $C_{rand} = 0.00023$ corresponding to a random graph with the same size and same average degree. Another study considered a larger database made of 200 million documents collected from different crawls (Broder *et al.*, 2000). In this case we also find power-law degree distributions with exponents $\gamma^{out} = 2.72$ and $\gamma^{in} = 2.1$. The value of the diameter in this data set is around $d = 16$. If we do not consider the edge orientation, we find a smaller value of $d = 6.83$. A more detailed analysis of the nature of the hyperlinks allows us to sketch a possible shape for the structure (see next section). An analysis of the largest database available, made of about 300 million documents and nearly 2.1 billion edges, confirmed the previous results (Laura *et al.*, 2003). It is interesting to note that in these two most recent and largest data sets, the out-degree probability distribution $P^o(k^{out})$ is poorly fitted by power-law functions. A better fit could be obtained by superimposing an exponential cut-off, limiting the number of outgoing edges per page. This is not surprising by considering the intrinsic different nature of out and in-degree. The in-degree obtained by a page is the product of a process involving every user on the net. They can decide to put an edge in their page to a search engine or a news site. The effort of reaching let us say $10,000$ edges to the same page is equally shared by the millions of users. In the case of out-degree instead, the webmaster of one page has to manually or automatically provide a list of $10,000$ edges in the page to reach a similar value for k^{out}. Even if such large numbers of outgoing edges are theoretically possible, in practice for reasons of clarity, control, and readability of pages, they are extremely unlikely. This natural limit it is probably the origin of the finite cut-off observed.

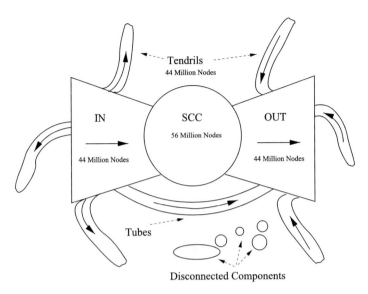

Fig. 9.11 The bow-tie picture of the WWW. The internal region of the strongly connected component is the only part where two random page have a direct path between them. Figure is adapted from (Broder *et al.*, 2000).

9.7.1   Small-world effect, the shape of web

Despite the small-world presence in sample of the Web (Albert, Jeong, and Barabási, 1999) it turns out that a couple of random pages can be also distant from each other. An explanation of this problem may arise by considering the large scale topology of the system. This is obtained through a representation of the Web (Broder *et al.*, 2000), by means of the so-called bow-tie picture (see Fig. 9.11). Considering the diverse nature of the directed edges, the WWW can be partitioned in different sets giving to it the shape of a bow-tie. In the strongly connected components (SCC) we have pages that can be reached from each other with a directed path. In the IN region we have all the pages from which we can reach the SCC, but which cannot be reached by pages in SCC. In the OUT region we have the pages reached from the SCC, but from which we cannot reach the SCC. In the tendrils there are pages reachable from the IN set or that can reach the OUT set. Tubes provide a direct connection between IN and OUT; finally disconnected components are isolated from the rest. Note that only 28 percent of pages (those in the SCC) experience the small-world phenomenon, this means that (unlike other networks) in this case if one takes two random vertices there is a high probability that they are disconnected.

	Italy	Indochina	WebBase	Altavista
n	$41.3M$	$7.4M$	$135.7M$	$203.5M$
m	$1.1G$	$194.1M$	$1.18G$	$1.46G$
SCC	$29.8M(72.3\%)$	$3.8M(51.4\%)$	$44.7M(33\%)$	$56.4M(27\%)$
In	$13.8K(0.03\%)$	$48.5K(0.66\%)$	$14.4M(11\%)$	$43.3M(21\%)$
Out	$11.4M(27.6\%)$	$3.4M(45.9\%)$	$53.3M(39\%)$	$43.1M(22\%)$
T	$6.4K(\ 0.01\%)$	$50.4K(0.66\%)$	$17.1M(13\%)$	$43.8M(22\%)$
D	$1.25K(< 0.01\%)$	$101.1K(1.36\%)$	$6.2M(4\%)$	$16.7M(8\%)$

Table 9.2 The values of order (n), size (m), Strongly connected component (SCC), In and Out Component, Tendrils (T), and Disconnected (D) for the various data sets.

This situation is continuously changing. The most recent studies (Donato *et al.*, 2005) extended these analyses to a crawl of nearly 360 million pages and to specific subgraphs of the web obtained by collecting pages under the national domains of Italy, Laos, Vietnam, and Cambodia (the latter grouped together in the 'Indochina' Web). While results on the large crawl confirm the previous results, it seems that geographical subsets display different statistical properties (see Table 9.2).

9.8   E-mail networks

The possibility of exchanging messages between users of the Internet was one of the first services provided, even before the introduction of the WWW. Since the advantages of this system of communication and the wide diffusion of computers, the users of the e-mail network are probably in the number of hundreds of millions or more. The most intuitive network representation of this system can be done by considering the various users as vertices and an exchange of e-mail as an edge. Unfortunately investigation on these edges is rather complicated. Track of an e-mail is usually recorded in the *log file* of the servers that set up the communication between the computer of sender and computer of the recipient. For privacy reasons (unlike hyperlinks in WWW) this is not public information. For that reason the study of e-mail networks has been concentrated in the analysis of some small subgraphs obtained by considering the traffic of e-mail only on specific servers of some companies or universities. The e-mail network can also be described by a directed multigraph where different edges represent different messages. The network is directed since not necessarily (but often) an e-mail from vertex

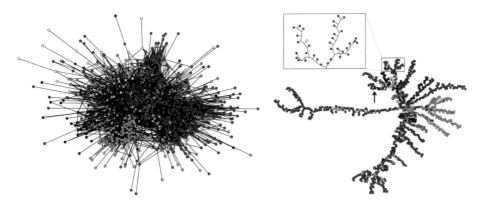

Fig. 9.12 Left: the network of e-mails in the University of Tarragona, different colours correspond to different departments. Right: the dendrogram obtained with the Newman Girvan algorithm of edge removal. Adapted from Guimerà *et al.* (2002).

i to vertex j does not imply an e-mail from vertex j to vertex i (Caldarelli, Coccetti, and De Los Rios, 2004b).

The two works we are describing here report the statistics obtained from the servers of the German University of Kiel (Ebel, Mielsch, and Bornholdt, 2002) and the Spanish University 'Rovira i Virigili' in Tarragona (Guimerà *et al.*, 2002). The network of the e-mails of the German university consisted of $58,912$ vertices forming a giant cluster of $56,969$ vertices and other disconnected components. This network shown in Fig. 9.12 is composed of addresses internal and external to the institution and was studied both in the directed and undirected version. In any situation we find the usual power-law behaviour $P(k) \propto k^{-\gamma}$ for the degree k. For the non-directed network we have $\gamma \simeq 1.8$. This exponent reduces a little bit by restricting the network to addresses internal to the Kiel domain. In this case we have $\gamma \simeq 1.3$. Finally also the directed network displays power laws in both in- and out-degree distributions. A somewhat different approach was taken in the case of the network of e-mails in the Spanish University of Tarragona (Guimerà *et al.*, 2002). The network was done by considering only the messages exchanged by about $1,700$ people working in the university. An edge was drawn when there was a *reciprocal exchange* between two users. Any message from bulletin boards and distribution lists was removed from the log files and only 'ordinary' e-mails were considered. The sum of these two effects, the bound on the local addresses and the non-personal messages removal, is probably responsible for the exponential distributions found for the degree. By applying the techniques of community detection based on the removal of edges with largest betweenness it is easy to verify that most of the traffic is due to acquaintances relations. Finally, it can be shown that the e-mail traffic develops self-organized coherent structures that are originated by synchronization between users(Eckmann, Moses, and Sergi, 2004). The different dynamical properties allow to distinguish organizational units (i.e. Departments) as the most static ones.

10. Social and cognitive networks

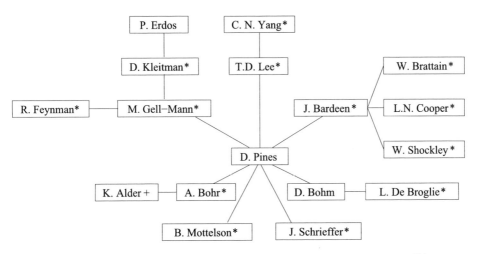

Fig. 10.1 The map of co-authorship for Nobel laureates in Physics (*) and in Chemistry (+) (after the on-line extended version of De Castro and Grossman (1999)).

When Professor Paul Erdős of Budapest announced that he would shortly be celebrating his eighty-first birthday with a party of special magnificence, there was much talk and excitement in the mathematical community. The special reason for the excitement was that this was probably the last time he could celebrate a birthday with a number of years that was a perfect square (Cerabino, 1994). Paul Erdős was not only the developer of graph theory, but also one of the most respected mathematicians of his time (Hoffmann, 1998). A great sign of distinction for his colleagues was to be numbered in the list of his collaborators and coauthors for a scientific paper. In his life he wrote many papers, with 507 distinct collaborators. These 507 are at a 'distance' 1 from Erdős. People collaborating with them (if not already listed in the 507 set) are at a distance 2 from Erdős. This 'Erdős Number' is then a measure of the proximity with such a genius.

This issue is so serious that various scientific papers have been written on this subject (De Castro and Grossman, 1999; Batagelj and Mrvar, 2000) and a web page (from which we took the above figures) is also devoted to the Erdős number (Grossman, 2003). Unfortunately, it is very difficult to be proud of a small Erdős number. Either you did not write a scientific paper in your life and therefore your distance is infinite, or your Erdős number is small. I am not very good at mathematics (Gabrielli, 2000) but nevertheless (as far as I know) my Erdős number is as small as 3.[68]

The idea of using networks to characterize social systems was introduced by an experiment on the structure of social communities (Milgram, 1967; Travers and Milgram, 1969). An example of collaboration network is shown in Fig. 10.1 The purpose of the experiment was to measure the network of acquaintances in human society. The idea was to send a passport-like packet to a few hundred randomly selected individuals in Wichita, Kansas. They had to deliver them to one 'target' in Cambridge Massachusetts. The study was then repeated by selecting random person in Omaha, Nebraska with 'target' living in Sharon and working in Boston, Massachusetts. A constraint in the delivering process was that each person must send the packet only to someone whom they knew on a first-name basis (thinking they were more likely to know the target than themselves). To this end some information was provided on the targets (name, address, occupation). Since every participant was also requested to tear off a card (after recording certain demographic details about themselves) and mail it back to the organizers of the experiment, it was possible to track all the paths to destination. The results of the experiment were surprising. Most of the messages arrived at their destination and this happened in a very short number of passages. A similar experiment on a much larger scale[69] was conducted at Columbia University (Dodds et al., 2003) yielding similar results. This phenomenon described as 'six degrees of separation' (or 'small-world effect') became lately widely known being the title of a Hollywood movie (F. Schepisi (Director) and J. Guare (Writer), 1993).

The cases presented in this chapter range from the structure of collaboration between people to the propagation of diseases in epidemics. We also present some discussion on various language networks. In this field, people

---

[68] A. Capocci, R. Ferrer i Cancho, G. Caldarelli → R. Ferrer i Cancho, O. Riordan, B. Bollobas, → B. Bollobas, P. Erdős (18 joint papers, first in 1962).

[69] See the website http://smallworld.columbia.edu

distinguish between cognitive approaches to language and social approaches: cognitive approaches focus on the structure of knowledge inside an individual's head, while social approaches focus on patterns of communication and interaction among individuals. From this point of view, everything presented here is definitely 'cognitive' rather than 'social'. Another issue that can be referred to as cognitive research is that of the organization of knowledge as it can be measured for example in an encyclopedia (this is a topic that can also be treated in conjunction with taxonomy). The case study is that of Wikipedia, which many consider a technological network, rather than a cognitive one. The situation is then open, as is often the case, for new evolving interdisciplinary fields. Here we make this division only for the sake of presentation of the results.

> Here we present some studies on the structure of networks of human interaction. The most important example are the co-authorship networks, sexual networks, and therefore the propagation of epidemics (note that for epidemics we can also refer to propagation of computer viruses in a completely different field). Similar considerations also apply to the different fields of language and knowledge whose structure seem to display some similar characteristics.

10.1 Networks of scientific papers

10.1.1 Co-authorship networks

One of the most difficult features when considering a network of persons is to characterize the quantity and quality of interactions between different people. Consider for example friendship: it is very difficult, if not impossible, to determine the degree of acquaintance between two persons. Anyway, by restricting the problem there is a specific area where we can measure the strength of connections. This is the case of scientific collaboration. It is a fair assumption to believe that two people have something in common if they write a paper together. Therefore, in the spirit of the Erdős number, one can generalize the question and consider the statistical properties of the whole network of co-authorships in one or more fields. This approach has a long tradition and it is interesting to note that one of the first models of preferential attachment was introduced in bibliometrics (de Solla Price,

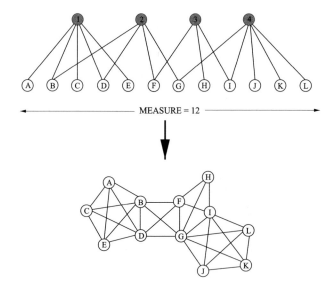

Fig. 10.2 The network of co-authorships is a bipartite graph. Dark vertices are the papers and white vertices authors. This network can be transformed into an author-only network by considering only the vertices in the bottom layer. We draw an edge between authors if they have written a paper together (they have an edge to the same dark vertex).

1965). The structure and features of collaboration networks are therefore one measure of the social interactions in this particular area. The structure of this particular network is given by the bipartite graph of the kind shown in Fig. 10.2. On one side there are papers, on the other the authors connected to their common papers. This bipartite structure can be transformed into a simple graph, by connecting to each other all the authors of the same paper. In this process we must expect a large global clustering, since the graph is made by little complete subgraphs connected to each other.

As regards the corpus of co-authorship networks, some extensive work has been done by collecting different data sets from electronic preprint databases (Newman 2001*a*, 2001*b*) and from other sets of data (Barabási *et al.* 2002, Newman 2001*c*, 2004*a*). The first database is formed by a corpus of 2, 163, 923 papers which can be extracted from the **Medline** database and whose topic is biomedical research. The second data set is a network of about 1.7 million papers from the database of **Mathematical Review** (Grossman and Ion, 1995). The third data set is composed of 156, 454 papers on Economics taken from the archive **Econlit**. This sample is particularly important since it allows us to investigate the dynamics of growth of the network. The data stored refer to papers published in the decades 1970–9,

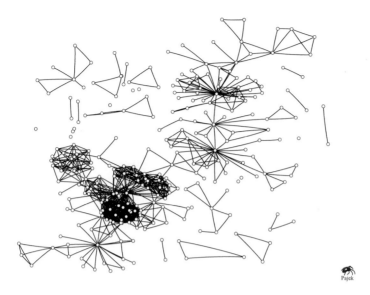

Fig. 10.3 The network of co-authorships in Santa Fe Institute. Different clusters correspond to different topics (Girvan and Newman, 2002). Data provided by M.E.J. Newman.

1980–9 and 1990–9 respectively (Goyal, van der Leij, and Moraga-Gonzàles, 2004). The fourth database is a corpus of $98,502$ papers submitted to the e-print archive at **http://www.arxiv.org**. This archive consists of different areas where authors can upload their contribution. These areas include, for example, condensed matter (cond-mat) and high energy physics (hep) (Newman 2001a, 2001b). The fifth in the series is composed of $66,652$ papers collected from the database **SPIRES**, hosting papers published in the area of high energy physics. The last data set collected consists of $13,169$ papers from **NCSTRL**, a preprint database in computer science.

Some authors may happen to be present in more than one data set even if the various archives correspond to rather different topics. These authors represent a sort of shortcut between different areas (as defined by editorial decisions of journals). This structure is particular evident from the network of collaboration in the Santa Fe Institute (an interdisciplinary research institution). In Fig. 10.3 we also see that in a small environment the communities of colleagues are visible (Girvan and Newman, 2002; Newman, 2004a). The main quantities characterizing these networks are reported in Table 10.1 In Fig. 10.4 we present the plot of the degree distribution (Newman, 2004a). Again we find distributions characterized by fat tails. In one case (the Medline archive) the function is well described by a power law. Another interesting quantity is that of the number of collaborations. As

	1	2	3	4	5	6
papers	$2,163,923$	$\simeq 1.75\ 10^6$	$156,454$	$98,502$	$66,652$	$13,169$
authors	$1,520,251$	$253,339$	$81,217$	$52,909$	$56,627$	$11,994$
papers/author	6.4	6.9	−	5.1	11.6	2.55
authors/paper	3.75	1.45	−	2.53	8.96	2.22
coll./author	18.1	3.9	1.67	9.7	173	3.59
g.c.	92%	82%	41%	85%	89%	57%
l	4.6	7.6	9.47	5.9	4.0	9.7
diameter	24	27	29	20	19	31
C	0.066	0.15	0.157	0.43	0.726	0.496
assortativity	0.13	0.12	−	0.36	−	−

Table 10.1 First, second and fourth columns are after Newman (2004a). The third column is after Goyal, van der Leij, and Moraga-Gonzàles (2004). The last two columns are after Newman, (2001a, 2001b). 1, 2, 3, 4, 5, 6 stand for the Medline, M. Review, EconLit, Physics, SPIRES, NCSTRL archives. The statistical error is on the last digit (or lower) of the figures. The size of giant component (g.c.) is the percentage with respect to the total size. When the network is not connected the average path length l is measured on the giant component. C indicates the clustering coefficient and assortativity is measured by means of assortativity coefficient. In this case it is computed as the assortativity coefficient defined in 1.3.3.

shown in Table 10.1 it ranges from the small values of 1.67 in economics and 3.9 in physics to the larger values of 18.1 in biomedicine and 173 in the high energy physics.

As regards the small-world effect, we find small values of the distance between authors. Typically, the path length is anti-correlated to the general clustering C. This is because the more the system is connected, the lower the distance between two vertices. Note the exception of computer science with a fairly large clustering and a large value of l.

A further analysis available only for Econlit is the evolution over time. For example, the average degree $\langle k \rangle$ changed from 0.894 in the 1970s to 1.244 in the 1980s to 1.672 in the 1990s. This can be related to an evolution in research habits but also to the substantial increase in the number of scientists in the field (increasing from $33,770$ in the 1970s to $81,217$ in the 1990s). The issue of evolution has also been treated considering measures of centrality to characterize the impact of authors (Börner et al., 2005). In this case a drift towards more cooperation in the production of scientific knowledge has also been observed.

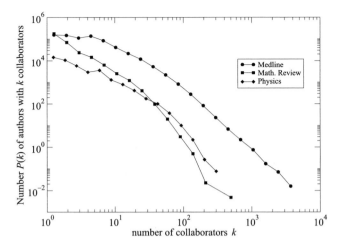

Fig. 10.4 Data on collaboration network provided by M.E.J. Newman (Newman, 2004*a*).

10.1.2   Citation network

Scientific papers form another kind of network when we consider the set of citations made between them . There are two main differences from the previous case. Firstly, the network is not bipartite; here the papers are the only kind of vertices connected by citations. Secondly, the papers are ordered according to the time of publication. In fact, a paper cannot cite papers that are not yet written (apart those in preparation, which are actually present even in a preliminary form) (see Fig. 10.5).

   A typical question in the field is to investigate if the inter-citation is based on real quality of cited papers or rather on personal knowledge. In other words, we want to know if the reason for the edge is more scientific than social. The case study presented here (White, Wellmann, and Nazer, 2004) investigates the inter-citation patterns of sixteen members of a research group on human development established in 1993. In this group of scientists there are 240 possible edges (the network is directed and therefore we have that this number is given by $n \times (n - 1)$ where n is the number of vertices). Half of the possible couples of scientists have a strong collaboration and 74 percent consider themselves friends or colleagues. Also in this case the inter-citation patterns are divided into four snapshots: prior to 1989, from 1989 to 1992, 1993 to 1996, and 1997 to 2000. Co-citation is shown to predict inter- citation; one cites those with whom one is co-cited. As members became better acquainted, citation of one another increased. Inter-citation was not randomly distributed, with a core group of twelve pairs predominating.

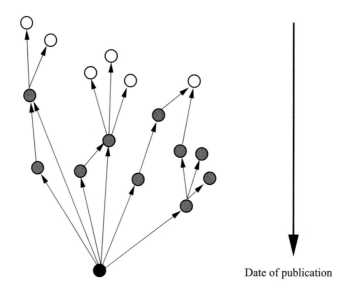

Date of publication

Fig. 10.5 An example of a very schematic citation network. Starting from one paper (black vertex) we consider all the old papers in the reference list. For every one of these we iterate the procedure. The colour of vertices in this picture is loosely related to the time of publication passing from dark (recent) to light (old). This procedure could in principle produce an exponentially large number of ancestors. Actually, since the corpus of all the scientific production is large but finite, we must expect that most of the ancestors coincide (i.e. strong presence of cycles in the graph).

Friends cited friends more than acquaintances, and inter-citers communicated more than non-inter-citers. However, intellectual affinity, as shown by co-citation, rather than social ties, leads to inter-citation.

The same database SPIRES considered above has also been analysed as a network of citations (Lehmann, Lautrup, and Jackson, 2003). The first result is that the probability that a given paper in the SPIRES database has k citations is well described by simple power laws. We have that $P(k) \propto k^{-\gamma}$, with $\gamma \approx 1.2$ for k less than fifty citations and $\gamma \approx 2.3$ for fifty or more citations. Different models reproduce these data. One of them generates power laws, while another generates a stretched exponential. It is not possible to discriminate between these models on the present empirical basis. A classification of citations distribution by subfield shows that the citation patterns of high energy physics form a remarkably homogeneous network. Furthermore, the extreme improbability that the citation records of selected individuals and institutions have been obtained by a random procedure can be demonstrated.

Another way to look at the same structure is to consider the corpus of citations to a single person (Lahèrrere and Sornette, 1998) or to a single paper (Redner, 1998). In the first case by considering the top scientists over the period 1981-97, one finds a stretched exponential distribution for the number of citations received. In the second case the statistics of citations are represented by a power law with exponent $-\gamma = -3$. This behaviour has been confirmed by successive analysis (Tsallis and de Albuquerque, 1999). From a different perspective, it is also possible to study the out-degree (Vázquez, 2001) that is the number of papers cited within a single paper. In this case the analysis reveals that at least for large values of the number of outgoing citations decays exponentially fast with two slopes characterizing two classes of journals, one with limitation on the page number (and therefore on the number of citations) and the second with no limitation. In the latter case the tail of the distribution is not very large, maybe because the number of papers that can be cited is in any case bounded by author knowledge.

The number and quality of citations also plays a role in the assessment of the success of scientific journals. In first approximation one can determine the *impact factor* of a journal with the ratio C_{it}/I where I is the number of papers published in a certain period and C_{it} is the number of citations these papers received. Put in this way the impact factor is nothing more than the total in-degree of the journal, normalized with the total number of publications. In the same spirit it is also possible to apply the Google PageRank algorithm (see Section 9.6.2) to assess the relative importance of one manuscript (Chen *et al.*, 2007). This study was done on the journal *'Physical Review'* in the period 1893-2003. While in general PageRank is correlated with the number of citations, some manuscripts that are outside this relationship (i.e. large PageRank and few citations) correspond to very famous and important pieces of research. Therefore it would be sensible to consider another way to measure the impact factor based on PageRank (Bollen, Rodriguez, and Van de Sompel, 2006).

10.2   Contact networks

10.2.1   Sexual networks

As already pointed out in the introduction, social networks are very difficult to define and measure when the vertices are individuals and the edges relationships amongst them. Degree of acquaintance, friendships or even love cannot be measured. Even if this were possible, the relations are not reciprocal; therefore the edges have one value for the first person and another for the second one. This difference can be so large that even the existence of

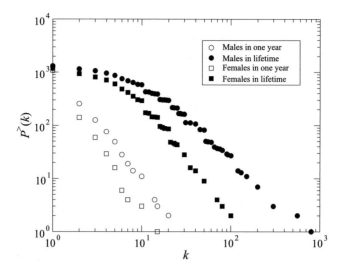

Fig. 10.6 Frequency distribution of the number of sexual contacts. Data provided by F. Liljeros (Liljeros *et al.*, 2001).

one connection can be questioned by one of the two persons concerned. In the case presented here (Liljeros *et al.*, 2001), we are dealing with a social network where this basic ambiguity is removed. The case is that of sexual contacts in a sample of the Swedish population. These data were collected in a Swedish survey of 1996 (Lewin, 1998) from a sample of 4,781 persons aged between 18 and 74 (of these only 2,810 persons (59%) answered). Because of the nature of the sexual intercourse, this network is highly dynamical and edges appear and disappear as relations are initiated and terminated. Averaging out the fluctuations we have the cumulative degree distribution shown in Fig. 10.6.

This degree distribution also has a power-law shape. We consider in the next section what implications this might have for this specific case. Here we want to extract some more information from a non evident feature of the plot in Fig. 10.6. If you note this plot, the area below the circles (males) and the squares (females) should be the roughly the same; instead the two areas are rather different. It has already been argued (Laumann *et al.*, 1994) that males tend to declare a larger number of intercourses than females. It is worth noting that 'proportional cheating' i.e. declaring a number proportional to the real one, does not change the shape of the distributions.

By reasoning in terms of absolute quantities the spread of the number of partners ranges from one to the thousands. These numbers at least for the order of magnitude are not strange and have been documented both in real life and in art.[70] A very simplified model that is nevertheless able to reproduce most of the above statistical properties is based on only one parameter (the 'collision rate') that drives the probability and the strength of interaction between two individuals (González, Lind, and Herrmann, 2006). The presence of such hubs can play a role in the way diseases spread out in a community as we shall see in the next subsection.

10.2.2   Epidemics

The study of epidemics is based on the fact that some diseases (as well as computer viruses) survive by passing from one host to another. Their diffusion is then connected with the structure of the individuals (or computers) that could be infected. We have already seen (as in the preceding subsection) that many social structures are scale-free networks with small-world properties. This fact plays a crucial role in the spread of epidemics and in the policies adopted to stop this diffusion. Because of the importance of the topic of disease control, much effort has been devoted both to data collection and to theoretical analysis (modelling). Many theoretical studies on infection spread are based on the assumption that the network of individuals is formed by a regular lattice of connections. This is not completely true (as for the sexual network) and a change of this basic hypothesis has important consequences for the behaviour of disease propagation (Pastor-Satorras and Vespignani, 2002).

The two basic models for infection propagation describe two different situations (Diekmann and Heesterbeek, 2000; Anderson and May, 1992). The first, is called **SIR** after the various possible states of an individual (Susceptible, Immune, and Recovered). This model describes diseases for which you have permanent immunization after you recover from them. The second one is called **SIS**, after the initial of Susceptible, Immune, and Susceptible and describes all the diseases for which immunization is not permanent. This means that you can again suffer the same infection (e.g. cold but also most computer viruses). In both cases a scale-free network of connections has the effect of dramatically increasing the strength of the infection.

Let us focus on the standard description of the SIS model; one key parameter of the model is the spreading rate λ. This is given by the ratio of

---

[70] Consider the famous Aria nr.4 'Madamina il catalogo è questo' by Lorenzo Da Ponte in the libretto for Mozart's *'Don Giovanni'* (Da Ponte and Mozart, 1787). 'In Italia seicento e quaranta, in Lamagna duecento e trent'una, cento in Francia, in Turchia novant'una, ma in Ispagna son giá mille e tre' (six hundred,forty in Italy, two hundred,thirty one in Germany, one hundred in France, ninety one in Turkey, but in Spain are already one thousand and three). This list accounts for $2,065$ partnerships; the same order of magnitude as the tail of the above distribution.

the probability ν of a vertex being infected divided by the probability δ of recovering from the disease, that is $\lambda = \frac{\nu}{\delta}$. The average number of connections for an individual is given by the quantity $\langle k \rangle$ (the degree of the vertex). One can write an equation for the evolution in time of the percentage $\rho(t)$ of infected people. There are two contributions to consider. Firstly, at time t all the individuals affected at time $t-1$ are now recovered. Secondly, those who were healthy at time $t-1$ but neighbours of infected ones can develop the disease. At the steady state these two contributions must balance and the value of $\rho(t)$ is constant.

 As a first approximation we can write

$$\frac{d\rho(t)}{dt} = -\rho(t) + \lambda\langle k \rangle \rho(t)(1 - \rho(t)). \qquad (10.241)$$

The second term of right-hand side is composed of the probability of being healthy $(1-\rho)$, times the probability of having a neighbour infected, that is $\langle k \rangle \rho$. The above quantity is then multiplied by the spreading rate λ. In this expression all the individuals have the same number of connections and furthermore we assume that the strength of the infection is proportional to $\rho(t)$. This hypothesis that corresponds to the 'mean field' approach in physics is known as homogeneous mixing (Anderson and May, 1992) in epidemics. By making the left-hand side equal to zero and indicating as ρ the stationary value for $\rho(t)$ we have that

$$\rho = \lambda\langle k \rangle(1 - \rho)\rho. \qquad (10.242)$$

The requirement of a characteristic number of connections defines an epidemic threshold $\lambda_c = \langle k \rangle^{-1}$ that we can use to compute the value of ρ. We obtain

$$\rho = 0 \qquad \qquad \text{if} \quad \lambda < \lambda_c$$
$$\rho = (\lambda - \lambda_c)/\lambda \qquad \text{if} \quad \lambda \geq \lambda_c. \qquad (10.243)$$

The plot of the variation of this quantity with respect to λ constitutes the phase diagram of the system, as shown in Fig. 10.7. From the above formula we see that this model predicts a *non-zero epidemic threshold*. If the spreading rate is larger than the threshold, the infection spreads and become persistent. Instead below the threshold the infection dies out. In the language of physics this is an absorbing state. The main hypothesis behind these results is the homogeneity of the number of connections; that is $\langle k \rangle \simeq$ constant.

In the case of a scale-free network the above statements are no longer correct. The same equations must now consider the case of non-homogeneous

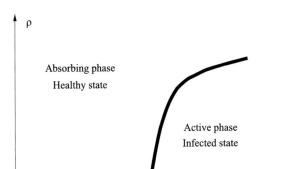

Fig. 10.7 The plot of the states of the system obtained by varying the percentage ρ of individuals infected and the spread of the disease λ.

distribution of infected individuals. This corresponds to introducing a function ρ_k of infected nodes with degree k and a function $\Theta(\rho_k)$ that gives the probability that any edge points to an infected node.

☢ The dynamical equation for the quantity ρ_k becomes

$$\frac{d\rho_k(t)}{dt} = -\rho_k(t) + \lambda k[1 - \rho_k(t)]\Theta(\rho_k(t)). \tag{10.244}$$

The assumption is that Θ is a function of the partial densities of infected nodes. In the steady (endemic) state, the latter are functions only of the spreading rate. Thus, the probability Θ also becomes an implicit function of the spreading rate, and by imposing the stationarity condition one finds

$$\rho_k(t) = \frac{k\lambda\Theta(\lambda)}{1 + k\lambda\Theta(\lambda)}. \tag{10.245}$$

The simplest form for Θ can be obtained by imposing no correlations between the degree of various nodes, on average the probability that an edge is drawn to a vertex whose degree is s is simply $sP(s)/\langle k\rangle$. Therefore, we obtain

$$\Theta(\lambda) = \frac{1}{\langle k\rangle}\sum_{k'} k'P(k')\rho_{k'} \tag{10.246}$$

in which, by putting the expression given by eqn 10.245, we have

$$\Theta(\lambda) = \frac{1}{\langle k\rangle}\sum_{k'} k'P(k')\frac{\lambda k'\Theta}{1 + \lambda k'\Theta}. \tag{10.247}$$

The trivial solution of this equation is $\Theta = 0$. But a non-zero stationary prevalence ($\rho_k \neq 0$) is also ensured when the following inequality is satisfied

$$\frac{d}{d\Theta} \left(\frac{1}{\langle k \rangle} \sum_{k'} k' P(k') \frac{\lambda k \Theta}{1 + \lambda k \Theta} \right) |_{\Theta=0} \geq 1. \tag{10.248}$$

The value of λ for which the left-hand side of the above equation gives 1 defines the value of λ_c.

Taking into consideration the above ingredients one finds that now the value of the epidemic threshold is given by

$$\lambda_c = \frac{\langle k \rangle}{\langle k^2 \rangle}. \tag{10.249}$$

This means that whenever the denominator goes to infinity (as happens for scale-free networks whose degree exponent γ is such that $2 < \gamma < 3$) we have $\lambda_c = 0$.

10.3   Linguistic networks

Linguistic networks in a loose sense are all those networks that can be established by studying the features of a particular (or more than one) language.

Most of these are composed of words related to each other by some kind of relationship related to position, syntax, or grammar. The simplest possibility is to start from a text document; after that one considers all the words present as the vertices of the graph and draws an edge between them according to the relationship considered. In almost all cases this results in a scale-free degree frequency distribution. One possible explanation of this behaviour could be the presence of Zipf's law. This law states that the frequency of words in a text follows a power-law distribution.

10.3.1   Zipf's law. The word frequency distribution

One elementary model reproducing Zipf's law is related to the exponential combination already seen in Section 4.3.2. Consider that words are randomly formed by collecting letters randomly. A word is complete when we select as 'letter' a separation space. All the letters are equiprobable, so that if the probability of having the separation space at any letter extraction is p, and the letters are m, their probability is $q = (1 - p)/m$. The probability x with which a particular word of y letter is extracted is given by

$$x = \left[\frac{1-p}{m}\right]^y p \propto e^{\mu y}, \qquad (10.250)$$

where $\mu = \ln(1-p) - \ln(m)$.

The distribution of words of length m is an exponential distribution $p(y) = m^y$ or $p(y) = e^{y\ln(m)}$. Therefore the distribution of the x (i.e. the distribution of the frequency of the words) is a power law (Mandelbrot, 1953)

$$P(x) \propto x^{-\gamma} = x^{-(1-\frac{\alpha}{\mu})}. \qquad (10.251)$$

Since

$$\gamma = 1 - \frac{\alpha}{\mu} = \frac{2\ln(m) - \ln(1-p)}{\ln(m) - \ln(1-p)}, \qquad (10.252)$$

and in most cases m is large and p small we obtain $\gamma \simeq 2$, which is in good agreement with most of the data.

10.3.2 Co-occurrence networks

The most immediate relationship is given when words are simply neighbours in the text. This would mean that in the elementary sentence 'John eats pizza' the vertices are 'John', 'eats' and 'pizza' and the edges are between 'John' and 'eats' and between 'eats' and 'pizza'. This rule can be loosened a little bit and in general one can put an edge between words even if they are a little further apart. For linguistic reasons (Kaplan, 1955) it is possible to connect all the words within a distance of two[71] (i.e. the nearest neighbours and the next nearest neighbours) (Ferrer i Cancho and Solé, 2001). The first data set we present here is composed of the British National Corpus as registered on the website http://info.ox.ac.uk/bnc/. This consists of about 10^7 words that can be linked in various ways giving rise to different networks. The first is created by the above defined rule of drawing an edge between two words at a distance less than or equal to two. This case will be indicated by the name 'Unrestricted Word Network' (UWN). The second network is a subset of the first, where an edge is actually drawn only if the co-occurrence has a probability larger than in a random case. This case will be denoted as the 'Restricted Word Network' (RWN). Finally they define also a kernel word network (KWN) where the only words present are those of a 'kernel',

---

[71] Statistically 87 percent of the syntactic relations take place at this distance (Kaplan, 1955). This of course does not mean that all edges drawn at a distance less than two are syntactically meaningful. Rather roughly one half of the edges drawn do not correspond to a syntactically valid relation.

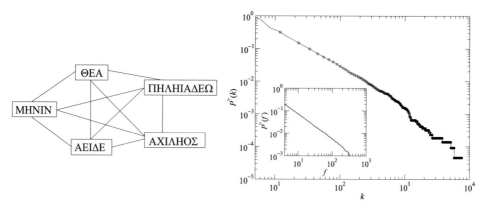

Fig. 10.8 An example of the network that can be drawn out of the first verse of Homer's Iliad (Sing, o Goddess the anger of Achilles son of Peleus). On the right the plot of the degree distribution for such a network. Note in the inset the plot of Zipf's law for this text.

	Size	*Measure*	*C*	$\langle d \rangle$	$\langle k \rangle$
Unrestricted (UWN)	478, 773	66, 652	0.687	2.63	74.2
Restricted (RWN)	460, 902	$1.77 10^7$	0.437	2.63	70.13
Kernel (KWN)	5, 000	$1.61 10^7$	–	2.63	1219

Table 10.2 The value of the graph quantities in the graph. Both UWN and RWN display a degree probability distribution.

that is all the words that are common to all the speakers of the language. The frequency of these common words follow a Zipf's law with an exponent different from the rarest ones. So the same authors claim that this set of words in the kernel is characterized by the change in the exponent in the Zipf's law. The striking features of all these networks, UWN, RWN and KWN is the presence of scale-invariance and small-world effect (see Tab. 10.2). A similar analysis also holds when considering poems. The source of data is the Iliad and as shown in Fig. 10.8 the words in the same verse are linked together. Remarkably, this network also shows evidence of scale-invariance in the degree probability function (Caldarelli and Servedio, 2006). This is very likely related with the Zipf's law for the same word as already shown for this case study in Fig. 3.17.

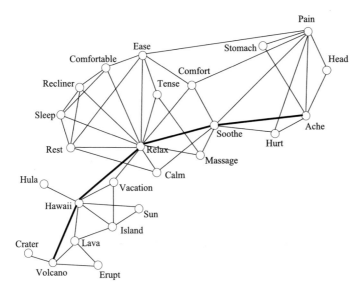

Fig. 10.9 A portion of the network of word association coming from 'Volcano' to 'Ache'. After Steyvers and Tenenbaum (2005).

10.3.3   Word associations

On top of the syntactic relations between the words, we can also define a network where the words are related by some more complicated relationship.

  One example of these linguistic networks is given by the *word association network*. In this case the structure is given by an experiment carried out as follows. A psychologist prepares a list of words ('stimulus') that is read to persons taking part in the experiment. The persons must respond with the first word that comes in their mind after the stimulus. These words are then used in a new experiment as stimuli and proceeding in this way a whole interconnected web is created (see Fig. 10.9). These patterns have been analysed from the point of view of their social importance (Steyvers and Tenenbaum, 2005) as well as a test case for an algorithm of community detections (Capocci *et al.*, 2005). The data set is formed by the results of one specific experiment with more than 6,000 participants and 5,000 initial words (Nelson, McEvoy, and Schreiber, 1998).

  Another way to put words together is by means of a *Thesaurus*. A thesaurus is a collection of synonymous (same meaning) or antonymous (opposite meaning) words. The idea is to connect two words if they happen to be synonymous. An extension of this concept is represented by Wordnet, where the network is composed by different classes of words. These two categories are called 'forms' and 'meanings'

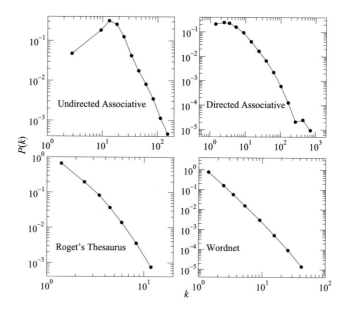

Fig. 10.10 The power law distribution of the various syntactic network. Data provided by M. Steyvers (Steyvers and Tenenbaum, 2005).

- An edge is drawn between many word 'forms' and a word 'meaning' if the former have the same meaning (i.e. if the word forms are synonymous)
- an edge is drawn between a word 'form' and the various word 'meanings' it can have (polysemic words)
- an edge is drawn between two words meanings if they are antonymous
- an edge is drawn between two words meanings if they have n hypernymous or meronymous relation[72].

In the literature two main corpus have been studied: The Merrian-Webster Dictionary and the Roget's Thesaurus (Roget, 2002). In these two cases of Wordnet and Thesaurus the two networks show remarkably features of scale invariance. In Fig. 10.10 we report the various degree distributions for such syntactic networks; after Steyvers and Tenenbaum (2005).

---

[72] A meronym is the name of a portion of something, or the material constituting something. 'Beer' is a meronym of 'drinks', 'Goalkeeper' is a meronym of 'football player' etc. An hypernym is a word whose meaning encompasses the meaning of another word.

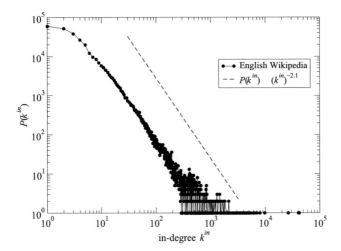

Fig. 10.11 The frequency distribution of the English version of Wikipedia.

10.4   Wikipedia

The last system we present here has something in common both with the cognitive networks with which we organize the meanings through language and with technological networks. The graph is that of Wikipedia (http://www.wikipedia.org) a free on-line encyclopedia. The whole system works on the same principle of hypertext markup language as the World Wide Web. That is, there are documents with mostly text, but also pictures, and other materials that are mutually accessible by means of the system of the hyperlinks. Every entry of the encyclopedia is a separate page and people can decide (and actually do) to connect different pages according to some kind of relation that might exist between the two entries. This system grows constantly as new entries are continuously added by users through the Internet. This is made possible by means of Wiki software. This software allows any user to introduce new entries and modify old ones. It is natural to represent this corpus of documents as a directed graph, where the vertices correspond to entries and edges to hyperlinks (autonomously drawn by the thousands of contributors). The distribution of both in-degree and out-degree is a power law (see Fig. 10.11). From the point of view of the medium used this is certainly a technological network. To be precise it is a thematic subset of the World Wide Web, since any entry is also a legitimate web page. On the other hand is the result of a cognitive process since entries are arranged (and that's a new part with respect to conventional encyclopedia) according to the perception of the readers. In this respect the system is also very similar to a word association network. One important feature of Wikipedia is that the users build different version of it in different languages. In principle, this could allow us to check if different cultures and ways of thinking result in different organization of the knowledge.

Some analysis has been done on the system (Holloway, Božičević, and Börner 2005; Capocci *et al.* 2006; Zlatić *et al.* 2006) showing most of the statistical properties of the system. The various languages form different graphs ranging from the English version of more than one million entries, to a few thousands for less common languages.

As a general remark the Wikipedia system is still expanding, new vertices are added continuously, and the number of edges is growing as a power of the number of vertices. In particular $m(t) \propto n(t)^{1.14}$. The Wikipedia shows a rather large interconnection; by using the bow-tie representation introduced for the WWW we find that most of the vertices are in the SCC. From almost any page it is possible to reach any other. For the English version (the largest) of the Wikipedia the in-degree exponent is $\gamma^{in} \simeq 2.1 \pm 0.1$ in almost all the analyses cited. For the out-degree instead, the various data sets analysed show a range of values from $\gamma^{out} \simeq 2.1$ to $\gamma^{out} \simeq 2.6$. A possible explanation for this discrepancy is given by the different times at which data sets were collected. No particular structure is present from the point of view of the clustering coefficient. In this case also different studies report slightly different result, in one case (Capocci *et al.*, 2006) no assortativity was measured; in another the assortative coefficient (defined in eqn 1.30) is very small ($r = -0.10 \pm 0.05$) (Zlatić *et al.*, 2006). Particularly important is the fact that different networks (in different languages) show very similar behaviour (Zlatić *et al.*, 2006). This suggests that the structure of the organization of knowledge might actually be similar in different cultures.

11. Financial networks

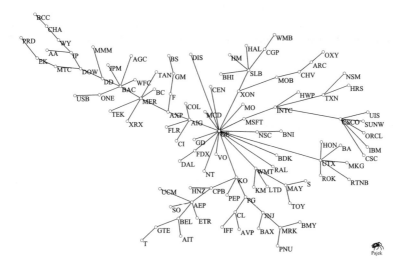

Fig. 11.1 A network of stock correlations in the New York Stock Exchange (Bonanno *et al.*, 2004). The various stocks are indicated by their symbol. The central vertex GE corresponds to General Electric.

In my younger and more vulnerable years my father gave me some advice that I've been turning over in my mind ever since. The advice is always to give credit when deserved. Even if the study of self-invariance is now very fashionable, it is necessary to say that the beginning of this activity traces back at least to Vilfredo Pareto (1897). The activity of this Italian economist is so important that even the shape of the power law is often known as a 'Pareto distribution' in the field of economics (see Section 3.6). Not surprisingly, one of the first studies of fractals started from the analysis of the oscillations of cotton prices in the commodity market (Mandelbrot, 1963). Those price oscillations showed a self-similar behaviour and could be described by a Lévy distribution[73] whose parameter α is an intrinsic measure of price volatility. The variance of this distribution is infinite. Since many features of economic systems (especially in finance) have been shown to possess a fractal behaviour it is not surprising that many financial networks also display scale-free properties.

Hitherto, various cases of financial networks have been considered. They are mostly related to the relation between different companies. This relation can take the form of a set of common directors on the board, participation in the capital of another company, a correlation in the stock prices (if they are quoted) as shown in Fig. 11.1, or lending of money if the companies are banks. In almost all the cases the interest in such analysis is also social, because those networks also describe interaction on a wider scale between individuals or groups of individuals with a common interest. This is particularly evident in the case of the board of directors where the knowledge of the topology has a certain importance in order to determine who controls what. If the same persons sit on the board of different companies, this creates an oligarchy. This causes a variety of problems from insider trading to lobbying against the interests of the many stock owners. A similar issue holds when considering mutual ownerships in traded companies. One company owning shares in another can extend its influence and proceed against the interests of the controlled company. Using companies as vertices one can define another type of network. In this case the edges are correlations between the stock prices in the market. When the companies are banks it is also possible to investigate the set of connections given by lending of money. In the same spirit the trade between different countries can be represented by means of a network. In all these cases graph theory can help in understanding more about the system and about the nature of interconnection amongst financial agents.

---

[73] A Lévy distribution is expressed through its Fourier transform $F[P_N(k)](x) = F[e^{-N|k|^\alpha}](x)$.

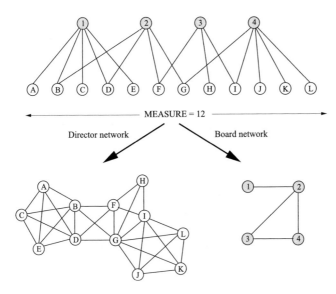

Fig. 11.2 As in Fig. 10.2 another process of formation of homogeneous networks from a bipartite one. In this case we investigate the structure obtained from both the layers.

In this chapter we will apply the formalism of graphs to different case studies in the area of economics and finance. Most of them refer to the activity of companies and range from the analysis of the board of directors, to the analysis of their portfolio (how many stocks of one company are owned by another) to their price correlation. When the companies are banks we also present the structure of the lending network formed by banks who constantly lend and borrow money from each other. Finally we present the weighted network of the trade between different countries.

11.1   Board of directors

This network is formed like that of co-authorships (see Fig. 11.2). Again in this case we have a bipartite graph of the kind shown in Fig. 10.2. On the upper layer we have companies and on the lower layer we have the people sitting on the boards of these companies. We can transform this bipartite graph into a graph where all the vertices are boards or all the vertices are

	B 86	B 02	B US	D 86	D 02	D US
N	221	240	916	2,378	1,906	7,680
M	1,295	636	3,321	23,603	12,815	55,437
Nc/N	0.97	0.82	0.87	0.92	0.84	0.89
$k/kc(\%)$	5.29	2.22	1.57	0.84	0.71	0.79
b/N	0.736	0.875	1.08	1.116	1.206	1.304
$\langle C \rangle$	0.356	0.318	0.376	0.899	0.915	0.884
$\langle d \rangle$	3.6	4.4	4.6	2.7	3.6	3.7
r	0.12	0.32	0.27	0.13	0.25	0.27

Table 11.1 The first three columns report data for the Boards (B) network in Italy in 1986 (B 86), in Italy in 2002 (B 02), and in USA in 1999 (B US). The last three columns report the data for the Directors (D) network in the same cases (1986, 2002 in Italy and USA in 1999).

directors. In the former case we connect all the boards sharing a common director (*boards network*). In the latter case (*directors network*) we connect all the people sitting on the same board.

Here we present the analysis of two different sets of companies, those put on the Fortune 1000 archive[74] and those quoted on the Italian stock exchange. The first data set consists of the 1,000 largest corporations in the year 1999 in the USA (Davis, Yoo, and Baker, 2003). Further analysis of the same data and a new set of data for the Italian stock exchange in the years 1986 and 2002 found similar results (Battiston and Catanzaro, 2004). All these networks present small-world properties, all of them are assortative, and they are also highly clustered. The presence of *lobbies* in boards turns out to be a macroscopic phenomenon in all data sets.

Whenever a director sits in more than one board we have an *interlock*. If this happens for more than one director, then we have a *multiple interlock*. This is the minimum structure of a *lobby* that is the subgraph of directors who happen to sit together on more than one board (Battiston, Bonabeau, and Weisbuch, 2003). The various properties of boards of directors are listed in Table 11.1.

In all cases the distribution of degree is fat-tailed and in some cases it can be fitted with a power law (see Fig. 11.3). The clustering coefficient also show interesting behaviour. When all the directors sit on only one board

---

[74] To download more recent data, the source is now the Fortune 500 with the 500 largest corporation. Presently the link to this archive is http://money.cnn.com/magazines/fortune/fortune500

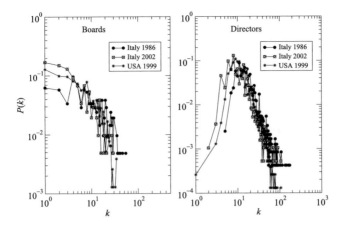

Fig. 11.3 Left: the plot of the cumulative degree distribution for the boards. Right the plot of the degree distribution for directors network. After Battiston and Catanzaro (2004).

their network is made of disjoint complete graphs (one for every board) of order n (where n is the number of directors in the specific board). In this case the average clustering coefficient C of the directors graph is one; conversely, the average clustering coefficient of the board graph is zero. Since instead some directors sit on more than one board we have substantial deviations from the above picture (even if C remains large for the directors network and small for the board one). All these values are reported in Table 11.1. The clustering coefficient is disassortative when plotted against the degree. The average of the neighbours degree (as shown in Fig. 11.4) displays a clear assortative behaviour.

11.2   Stock networks

11.2.1   Ownership networks

Another way to measure the influence of one company on another is to check if one of them owns a proportion of the other. This kind of network is called an 'ownership network' . The vertices are companies traded on a stock market and directed weighted edges indicate that one company owns a portion of another (the direction goes from the owned to the owner). The issue of ownership can be related to influence or control, but owning stocks of another company can also be a profitable way to invest a company's money. According to traditional portfolio theory (Markowitz, 1952)

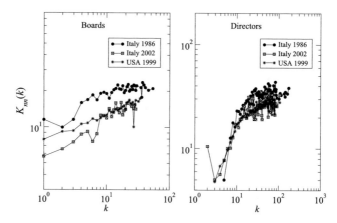

Fig. 11.4 Left: the plot of the average neighbour degree for the boards. Right: the plot of the same quantity for directors network. After Battiston and Catanzaro (2004).

one would like to invest the money under the double constraint of maximizing the return and of reducing the risk of the investment. A standard hypothesis in portfolio theory is that many assets are present in the market and anyone invests differently in them according to a personal strategy. Data analysis on stock ownership network shows a rather different result. In this case the number of investments is also characterized by power-law distributions, that is, many invest in few assets and few invest in many. This signals a significant deviation from the standard scenarios of portfolio optimization (Bouchaud and Potters, 2000) and allows us to characterize the different markets quantitatively.

The data sets analysed (Garlaschelli *et al.*, 2005) refer to the shareholders of all companies traded in three different markets. The 240 companies of the Italian stock market 'Borsa Italiana' (BI)[75] in the year 2002. The 2,053 companies traded in the New York Stock Exchange (NYSE) in the year 2000 and the 3,063 companies traded in the National Association of Security Dealers Automated Quotations system (NASDAQ)[76] in the year 2000. For every company only the largest investors are known (typically those with more than a few per cent of the total number of stocks). Some of these shareholders are individuals or companies not traded on the market; However, the investors are themselves often companies traded on the same market. In both cases this leads naturally to a network description

---

[75] Data from Banca Nazionale del Lavoro: La meridiana dell'investitore (2002).
[76] Data for the USA markets have been collected from http://www.lycos.com

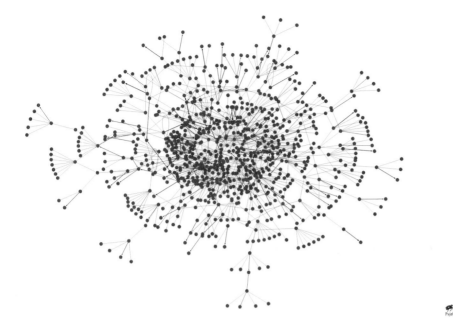

Fig. 11.5 The network of stock ownership in the Italian Stock Exchange.

of the whole system (see Fig. 11.5), where both investors and assets are represented as vertices and a directed edge is drawn from an asset to any of its shareholders. As we see in the following, the quantities describing the graph are different depending on whether or not we consider the external investors.

In this topological description the in-degree $k^{in}(i)$ (number of incoming edges) of company i corresponds to the number of different stocks in the portfolio. Vertices with zero in-degree are companies not investing in the market. The out-degree $k^{out}(j)$ of a vertex is the number of shareholders of the company j. The network is weighted since any edge is characterized by the amount s_{ij} of the shares of asset j held by i. This can be multiplied by the market capitalization $m_c(j)$ of the asset j transforming in monetary terms the weight of the edge. The quantity $w_i = \sum_j s_{ij} m_c(j)$ is therefore the *strength* of vertex i measured in money and it corresponds to the total wealth in the portfolio of i. As evident from Fig. 11.6 some of the degree distributions can be fitted with a power law. Since in the figure there is a cumulative distribution, this means that

$$P^>(k^{in}) \propto (k^{in})^{1-\gamma}. \tag{11.253}$$

The value of the exponent γ differs across the various markets: $\gamma_{NYSE} = 2.37$, $\gamma_{NASDAQ} = 2.22$, $\gamma_{BI} = 2.97$. This scale-invariance almost disappears

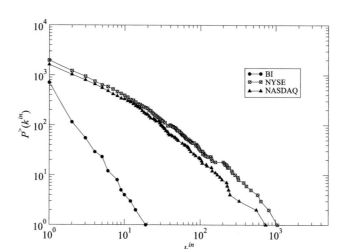

Fig. 11.6 The degree distribution for the NASDAQ, NYSE, and Italian Stock Exchange (BI).

when one restricts to the network of companies alone. That difference is less accentuated in the case of the Italian market. This might suggest that in the USA the market is dominated by large investors outside it, whereas in the Italian market the largest stock holders are the companies traded. For the small k^{in} region of $P^>(k^{in})$ the opposite occurs. This is reflected in the fact that only 7 per cent of companies quoted on US markets invest in other companies, while the corresponding fraction is 57 per cent in the Italian case.

Since this is a weighted network, we can also measure the distribution of the strength, that is the frequency $P^>(\omega)$ of investors whose portfolio volume is greater than or equal to ω. Once more (see Fig. 11.7 left) the tail of the distribution is well fitted by a power law

$$P^>(\omega) \propto \omega^{1-\alpha}, \tag{11.254}$$

corresponding to a probability density $\rho(\omega) \propto \omega^{-\alpha}$. The empirical values of the exponent are $\alpha_{Nyse} = 1.95$, $\alpha_{Nasdaq} = 2.09$, $\alpha_{BI} = 2.24$. Note that, since ω provides an estimate of the (invested) capital, the power-law behaviour can be directly related to the Pareto tails (Pareto, 1897; Badger, 1980; Drăgulescu and Yakovenko, 2001) describing how wealth is distributed in the economy. Consistently, the small ω range of $P^>(\omega)$ also seems to mimic the typical form displayed by the left part of many empirical wealth distributions. The characterization of this part is rather controversial (log-normal,

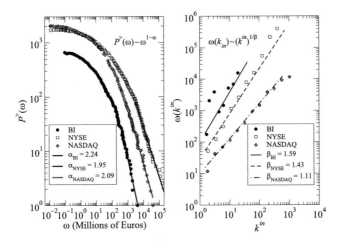

Fig. 11.7 Left: the frequency distribution of the money ω owned in the NASDAQ, NYSE, BI. Right: The correlation between in-degree and money.

exponential, and gamma distributions have all been proposed (Drăgulescu and Yakovenko, 2001) to reproduce it).

For this particular case, a fitness model of the kind presented in Section 5.6 is able to reproduce most of the data features. The main difference from the traditional formulation is that in this case we have two types of vertices in the network. They are the N investors (each characterized by its fitness x_i) and the M assets (characterized by a different quantity y_j). Note that sometimes a company can act as both types, in this case the two quantities coincide.

A natural choice is to consider x_i as proportional to the portfolio volume of i, which is the wealth that i decides to invest. The quantity y_j can instead be viewed as the information (such as the expected long-term dividends and profit streams) associated to the asset j. Generally the y_j can also be a vector of quantities (Boguñá and Pastor-Satorras, 2003). An edge is drawn from j to i with a probability which is a function $f(x_i, y_j)$ of the portfolio volume of the first one and the asset information of the latter one. Since the network is directed we have that $f(x, y) \neq f(y, x)$.

The simplest choice for the linking function is to use a factorizable form (Servedio, Caldarelli, and Buttà, 2004)

$$f(x, y) = g(x)h(y), \tag{11.255}$$

where $g(x)$ is an increasing function of x. This choice takes into account the fact that investors with larger capital can afford larger information and transaction costs. For that reason they are more likely to diversify their portfolios. The function $h(y)$ is related to the strategy used by the investors to process the information y relative to each asset. The stochastic nature of the model allows for two equally wealthy agents to make different choices even if assets with better expected long-term performance are statistically more likely to be chosen. For large networks, the expected in-degree of an investor with fitness x is given by

$$k^{in}(x) = g(x)h_T,$$ (11.256)

where h_T is the total value of $h(x)$ summed over all N agents.

 If the above relation is invertible, and if $\rho(x)$ denotes the statistical distributions of x computed over the N agents, then the in-degree distribution is given by

$$P(k^{in}) = \rho[x(k^{in})]dx(k^{in})/dk^{in}.$$ (11.257)

Analogous relations for $k^{out}(y)$ and $P(k^{out})$ can be obtained directly. However, since the information regarding k^{out} is incomplete one cannot test the model with respect to the function $h(y)$. The analogous choice for the attachment probability is then $g(x) = cx^{\beta}$ with $\beta > 0$, where c is a normalization constant ensuring $0 \le g(x) \le 1$ (a possible choice is $c = x_{max}^{-\beta}$, so that by defining $x \equiv w/w_{max}$ we can directly set $c = 1$). It is straightforward to show that the predicted expressions eqns 11.256 and 11.257 now read

$$k^{in}(x) \propto x^{\beta},$$ (11.258)

$$P(k^{in}) \propto (k^{in})^{(1-\alpha-\beta)/\beta},$$ (11.259)

where we have used the fact that $\rho(x) \propto x^{-\alpha}$ for large x.

Note that the above results still hold in the more general case when $f(x,y)$ is no longer factorizable, provided that $k^{in}(x) = M \int f(x,y)\sigma(y)dy \propto x^{\beta}$ as in eqn 11.258, where $\sigma(y)$ is the distribution of y computed on the M assets.

Note that we used a measure of the strength as fitness. Put in this way eqn 11.258 is exactly the observed relation reported in eqn 1.5 provided we have $\beta = 1/\zeta$.

One can show then that the empirical power-law forms of $\rho(x)$, $k^{in}(v)$, and $P(k^{in})$ are in qualitative agreement with the model predictions. Moreover, by comparing eqns 11.253 and 11.259 one can predict the following relation between the three exponents α, β, and γ:

$$\beta = (1-\alpha)/(1-\gamma).$$ (11.260)

By substituting in the above expression the empirical values of α and γ obtained through the fit of Fig. 11.6 and the first plot in Fig. 11.7, we obtain the values of β corresponding to the curves $\omega(k^{in}) \propto (k^{in})^{1/\beta}$ shown in Fig. 11.7, which simply represent the inverse of eqn 11.258 in terms of the quantity ω. Remarkably, the curves are all in excellent agreement with the empirical points shown in the same figure, except for some anomalous points for BI.

11.2.2   Stock price correlations

Another way to generate a network of companies is to consider how their stock prices are correlated. Whatever the mechanisms behind the formation of the stock price, it is common experience that companies operating in the same sector (e.g. financial or electronic) show similar price fluctuations. Using the various companies as vertices and connecting them according to the correlation in their prices we can define a network of stocks. Different authors have worked on this topic and defined similar networks of various sizes (i.e. number of edges). By keeping the minimal set of information one has that the resulting structure is a tree (Bonanno *et al.*, 2003). Other choices are possible (Onnela *et al.*, 2004; Tumminello *et al.*, 2005), for example, deciding in advance the number of edges.

The crucial variable is the daily price return $r_i(t)$ of company i on day t. From that, one can consider all the possible couples of companies and compute the correlation between the respective price returns. Two price stocks are correlated if they vary in a similar way. In other words companies i and j are correlated when the price of stock i increases if price of stock j also increases. Since we are interested in the variations in percentage (i.e. dr_i/r_i) we use a simple transformation. Using the properties of the logarithm the variation dr_i/r_i can be written as $d(\ln(r_i))$. For that reason people usually consider the logarithmic price return $p_i = \ln(r_i(t)) - \ln(r_i(t - \Delta t))$. Finally, we compute the correlation $\rho_{ij}(\Delta t)$ between these returns over a time Δt.

Correlation is computed by means of

$$\rho_{ij}(\Delta t) = \frac{\langle p_i p_j \rangle - \langle p_i \rangle \langle p_j \rangle}{\sqrt{(\langle p_i^2 \rangle - \langle p_i \rangle^2)(\langle p_j^2 \rangle - \langle p_j \rangle^2)}}. \tag{11.261}$$

This entry can be associated with a metric distance through the following relation

$$d_{ij} = \sqrt{2(1 - \rho_{ij})}, \tag{11.262}$$

between asset i and j through the relation (Gower, 1966; Mantegna, 1999).

Therefore the logical steps of the process are the following:

- we start from the price return $r_i(t)$ of a company i;
- we compute the logarithmic returns p_i over a time interval Δt;

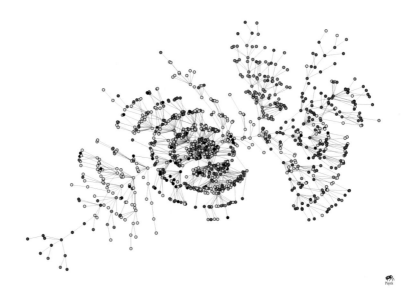

Fig. 11.8 Correlation based minimal spanning tree of real data from daily stock returns of 1,071 stocks for the twelve-year period $1987 - 98$ (3,030 trading days). The clustering of stocks corresponds roughly to Standard Industrial Classification System.

- we compute the correlations ρ_{ij} between the p_i of stock i and the p_j of stock j;
- we associate a distance d_{ij} related to correlation ρ_{ij}.

Different choices are possible: The first option is to build a minimal spanning tree (MST).[77] We repeat the method described in section 1.2.1. We start by ranking the distance from the shortest to the longest. We select the shortest distance and we connect the two vertices involved. We proceed along the list until we connect all the stocks. If by drawing an edge we create a cycle, we discard this distance and we take the next entry.

An alternative choice is simply to draw the desired number of edges irrespective of cycle creation or global connection of the graph.

11.2.2.1   Minimal spanning tree   We report here the analysis made for 1,071 stocks traded at the NYSE and continuously present in the twelve-year period 1987-98 (3,030 trading days) (Bonanno *et al.*, 2003). Fig. 11.8 shows

---

[77] The method of constructing the MST linking N objects is known in multivariate analysis as the nearest neighbour single linkage cluster algorithm (Mardia, Kent, and Bibby, 1980).

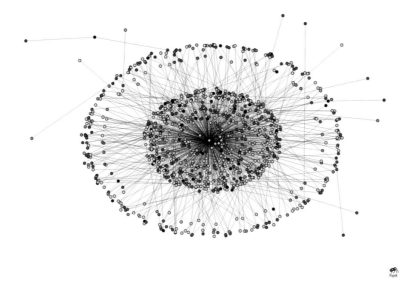

Fig. 11.9 Correlation-based minimal spanning tree of a numerical simulation of the one-factor model of eqn 11.263.

the network obtained from the real data. The clustering roughly corresponds to the division into the main industry sectors as provided by the Standard Industrial Classification system.[78] Regions corresponding to different sectors are clearly seen. For example, the mining sector stocks are observed to belong to two sub-sectors, one containing oil companies (located on the right-hand side of the figure) and one containing gold companies (left-hand side of the figure). It is interesting to note that this network representation allows us to validate models of price fluctuations. The simplest model used in this field assumes that the return time series are given by uncorrelated Gaussian time series, that is $r_i(t) = \epsilon_i(t)$, where $\epsilon_i(t)$ are Gaussian random variables with zero mean and unit variance. This type of model has been considered (Laloux *et al.*, 1999; Plerou *et al.*, 1999) as a null hypothesis in the study of the spectral properties of the correlation matrix. The one-factor model assumes that the return of assets is controlled by a single factor (or index). Specifically for any asset i we have

$$r_i(t) = \alpha_i + \beta_i r_M(t) + \epsilon_i(t), \tag{11.263}$$

[78] The Standard Industrial Classification system can be found at http://www.osha.gov/oshstats/naics-manual.html

where $r_i(t)$ and $r_M(t)$ are the return of the asset i and of the market factor at day t respectively, α_i and β_i are two real parameters, and $\epsilon_i(t)$ is a zero mean noise term characterized by a variance equal to $\sigma_{\epsilon_i}^2$. A possible choice is the Standard & Poor's 500 index, assuming also that $\epsilon_i = \sigma_{\epsilon_i} w$, where w is a random variable distributed according to a Gaussian distribution. The model parameters can be estimated for each asset from real time series with ordinary least squares method (Campbell, Lo, and MacKinlay, 1996) and those parameters are used to generate an artificial market according to eqn 11.263. The fraction of variance explained by the factor r_M is approximately described by an exponential distribution with a characteristic scale of about 0.16.

In the structure obtained with the random model, few nodes have a degree larger than a few units. This implies that the MST is composed of long files of nodes. These files join at nodes of connectivity equal to a few units. In Fig. 11.9 we show the MST obtained in a typical realization of the one-factor model performed with the control parameters obtained as described above. It is evident that the structure of sectors of Fig. 11.8 is not present here. Rather there is a star-like structure with a central node. The largest fraction of vertices points directly to the central one and a smaller fraction is composed by the next-nearest neighbours. Very few vertices are found at a distance of three edges from the central vertex.

The degree distribution for the real data does not show any clear behaviour. The highest degree $k_{max} = 115$ is observed for General Electric, one of the most capitalized companies on the NYSE. The value of the maximum degree is small, $k_{max} = 7.34 \pm 0.92$, showing that no asset plays a central role in the MST.

11.2.2.2   Non-spanning trees In the second case we can start from the set of vertices and insert the edges one by one according to the rank of the distance, so that the first two stocks are those less distant or more correlated (Onnela et al. 2003; Onnela, Kaski, and Kertész 2004). Financial data sets analyzed refer to a set of 477 stocks traded continuously from 1980 to 2000 in NYSE. In this case it is also possible to obtain a structure with $n - 1$ edges, which would not necessarily correspond to a tree. Since spanning the network is no longer an issue, the set of edges will very likely connect clusters of stocks. This is exactly what happens in the right-hand picture of Fig. 11.10. Here the number of edges is larger than that of a tree and still some vertices are isolated and remain isolated also by adding new edges. This is because the strength of correlation of those stocks with the others is rather low. Therefore while a spanning tree can inform us about the taxonomy of a market, the graph asset built in this way seems to represent

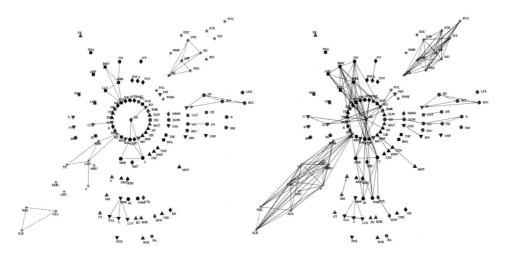

Fig. 11.10 Left: sample graph for $n = 40$ edges. Right: sample graph for $n = 160$ edges. After Onnela, Kaski, and Kertész (2004).

better the strength of correlations. As for the other general features of this asset graph, one can note that the process of adding new edges, those closing a cycle will appear earlier in the experimental graph than in a random graph. The total number of clusters also grows much faster with time here than in a random graph, signalling the presence of a strong correlation between some vertices. This correlation is so strong that it destroys the usual phase transition present in a random graph. In a random graph by continuously adding edges at a certain threshold all the vertices coalesce forming a giant cluster. This threshold is not present at all in the asset graph.

11.3   Bank networks

Another example presented in this chapter is that of the inter-bank market. In this case, the different banks operating in a certain market can be represented as a network. Given a pair of banks i and j a directed edge is drawn from i to j if bank i borrows liquidity from bank j. Money is exchanged between various private banks in order to make it available to the possible requests of bank clients. If requests of customers should exceed the liquidity reserve of a bank, this liquidity shock is absorbed by the market. The bank can then provide in real time the liquidity needed to operate. This is because central banks require banks to reserve a part of their deposits

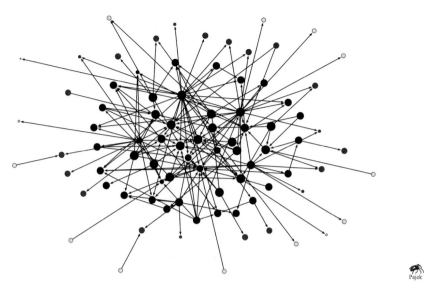

Fig. 11.11 A typical trading day. Grey level and size of vertices are proportional to transaction volumes of the banks. Small banks have light grey and small size. Dark grey and large sizes correspond to 'large banks'.

and debts with them, to create a buffer against liquidity shortages. In this sense central banks ensure the stability of the banking system avoiding liquidity shocks. This problem traditionally faced by economists and policy makers became more and more important after the introduction of the Euro money in some countries of the European Union and enlargement of national markets to a unique European market.

The particular case study presented here is that of the Italian market, which is kept under control by the European Central Bank (ECB). A special requirement of the ECB is that every bank must deposit on average the 2 per cent. of the total amount of their transactions, this percentage being calculated on the 23rd each month (hereafter this day will be indicated as end of month EOM). Given this requirement, the banks operate in order to maximize their return and to fulfil the above request. As appears clearly from the graph structure (De Masi, Iori, and Caldarelli, 2006), the various banks operating in a market are divided into different groups roughly related to their size. It is possible to classify the participating banks according to the volume of their transaction. Groups $1, 2, 3, 4$ have volumes in the range 0-100, 100-200, 200-500, over 500 million Euros per day respectively. Smaller banks make fewer operations and on average lend or borrow money from larger banks. The latter have many connections each other making many transactions, while the former do not. This feature does not change

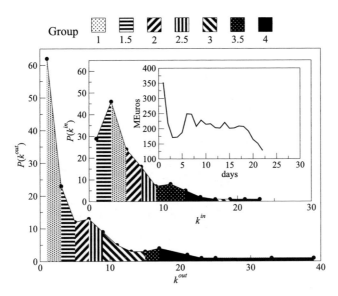

Fig. 11.12 A plot of the in- and out-degree distributions. As already noticed the contribution to the tail of frequency distribution comes from the banks of group 4. Here on top of the usual representation of the frequency distribution $P(k)$ for a certain degree k, there is a portion of the plot with a pattern related to the average group of the vertices. For example the average group of banks with degree 10 is of banks of group 3 (dark grey). For non-integer value of this average we introduced intermediate patterns.

very much in time, instead the volume does so, showing a rarefaction of transactions over time. The data set consists of all the overnight transactions concluded from January 1999 to December 2002 for a total of 586,007 transactions. When a trading day is represented as a network we immediately note, as shown in Fig. 11.11, that the core of the structure is composed by the banks of the last group.

There is also a tendency for banks of groups 1-2 to be mostly lenders. Banks of groups 3-4 which make more transactions, are more likely to be borrowers. This is particularly clear from Fig. 11.12 where the in-degree frequency distribution (money lent) and out-degree frequency distribution (money borrowed) in the network are presented. It is possible to compute the group of the banks whose degree is k and we represent this information by assigning a scale of grey patterns to the plot accordingly. In addition to this scale there are also some intermediate grey levels to account for intermediate values of the average between one group and another. The tail of the two distributions is black, i.e. it is made of banks of group 4. Those

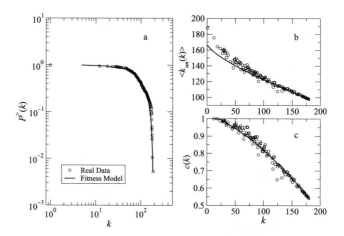

Fig. 11.13 Topological properties of the undirected WTW in the year 1995. (a) Cumulative degree distribution $P^>(k)$. (b) Plot of $\langle k_{nn} \rangle$ versus k. (c) Plot of $c(k)$ versus k. Circles represent empirical data, while the solid curves are the predicted trends derived from the fitness model.

tails are fatter in the case of the out-degree, signalling the tendency of large banks to be mainly borrowers. All the above quantities can be reproduced by means of a suitable model of network growth. The idea is that the vertices representing the banks are defined by means of an intrinsic character that could, for example, be their size. Edges are then drawn with a probability proportional to the sizes of the vertices involved.

11.4   The world trade web

The final example of this chapter is the network formed by the import/export trade relationships among all world countries, or *world trade web* (WTW in the following). Each world country is represented as a vertex $i = 1, \ldots, n$ and a trade relation (either imports or exports) between i and j is represented as an edge connecting i and j. This network is a picture of the macro-economy at a global scale, where the degree of a country i is the number of its trade partners (i.e. countries either importing from or exporting to i) (Serrano and Boguñá, 2003).

Several properties of this network have been studied (Li, Jin, and Chen, 2003; Garlaschelli and Loffredo, 2005). We first report a series of results regarding a snapshot of the WTW in the year 1995 (Garlaschelli and Loffredo, 2004a). The degree distribution $P(k)$ was found to display a very narrow power-law region followed by a sharp cut-off corresponding to a large number of countries having degree close to the maximum value n-1. This network is therefore not scale-free (Garlaschelli and Loffredo, 2004a) (see Fig. 11.13 a). The average nearest neighbour degree $\langle k_{nn}(k)\rangle$ versus the degree k gives the plot shown in Fig. 11.13 b. The decreasing trend signals that the WTW is a *disassortative* network displaying anti-correlation between the degrees of neighbour vertices. Therefore countries with many trade partners are on average connected to countries with a few partners. Finally, the clustering coefficient $c(k)$ of vertices whose degree is k is shown in Fig. 11.13 c. The decreasing trend signals in this case that the partners of well-connected countries are less interconnected than those of poorly connected ones.

The WTW is one of the few systems whose topological properties can be fully reproduced by one of the models described in Chapter 5. In particular, a fitness model (Caldarelli *et al.*, 2002) where *the fitness variable is identified with the gross domestic product (GDP) of each country* produces a network with the same properties of the WTW. In more detail, firstly we take the fitness x_i as the rescaled variable

$$x_i \equiv \frac{w_i}{\sum_{j=1}^{n} w_j} \tag{11.264}$$

where w_i is the empirical GDP of country i. Secondly, we choose a linking function $f(x_i, x_j)$ of the kind:

$$f(x_i, x_j) = \frac{z x_i x_j}{1 + z x_i x_j} \tag{11.265}$$

where z is the only free parameter of the model. This choice corresponds to establishing a trading channel with a probability proportional to the GDP of the countries involved. A natural choice for fixing the parameter z is to require that the expected number of edges $\bar{m} = \sum_{i<j} f(x_i, x_j)$ equals the observed number m. Once z is fixed, all the expected topological properties can be computed in terms of $f(x_i, x_j)$. For instance, the expected degree of vertex i is (see eqn 5.172)

$$\bar{k}_i = \sum_j f(x_i, x_j) \tag{11.266}$$

and the expected average nearest neighbour degree of i is (see eqn 5.174)

$$\langle \bar{k}_{nn} \rangle = \frac{\sum_j f(x_i, x_j) k_j}{\bar{k}_i} = \frac{\sum_{jk} f(x_i, x_j) f(x_j, x_k)}{\sum_j f(x_i, x_j)}. \tag{11.267}$$

Finally, the expected clustering of i is given by (see eqn 5.175)

$$\bar{c} = \frac{\sum_{jk} f(x_i, x_j) f(x_j, x_k) f(x_k, x_i)}{\sum_{jk} f(x_i, x_j) f(x_i, x_k)}. \tag{11.268}$$

All the above predictions of the model can be tested against the real data (Garlaschelli and Loffredo, 2004*a*). This is shown in Fig. 11.13. Remarkably, the model is in excellent agreement with the empirical data, so the fitness model captures the topological organization of the WTW successfully.

Note that the specification of the topology of the WTW requires in principle knowledge of $n(n - 1)/2$ parameters representing the entries of the corresponding adjacency matrix (the trade fluxes). Interestingly, the above model provides full knowledge of the expected properties of the network by exploiting only N parameters corresponding to the values of the GDP of world countries. Also note that the two sources of data, the trade flows and the GDP values, are completely distinct. Therefore, there is the surprising possibility of reproducing the full network topology in terms of a distinct, much smaller data set.

The above results refer to a single snapshot of the WTW. Obviously, the structure of the WTW changes continuously in time due to variations in the number $N(t)$ of world countries and to the continuous rearrangement of its connections, and hence of the adjacency matrix. It has been shown (Garlaschelli and Loffredo, 2005) that all the above properties are robust in time since they are displayed by all the annual snapshots of the WTW during a long time interval. There is a value $z(t)$ for each year t that controls the topology of the network. The corresponding expected properties of the WTW are always in very good agreement with the observed ones. Also, it is possible to refine the description by considering the WTW as a directed graph, by drawing a directed edge from i to j if money flows from i to j, or in other words if j exports some goods to i. It was shown that, due to the peculiar way of occurrence of mutual edges in the WTW (Garlaschelli and Loffredo, 2004*b*), the full directed description can be easily recovered from the undirected one. The expected properties in the directed case are a straightforward generalization of the expressions presented above (Garlaschelli and Loffredo, 2005), and they are again in excellent agreement with the empirical ones.

PART III

APPENDICES

Appendix A.   Glossary

- **Adjacency matrix.** A matrix A of size $n \times n$ where n is the **order** of the graph. The element a_{ij} is not zero if an edge goes from i to j. The value of a_{ij} is the weight of the edge or in a **multigraph** the number of multiple edges between vertices i and j. If edges have no direction the matrix is symmetric, i.e. $a_{ij} = a_{ji}$.
- **Betweenness.** A measure of the importance of a **vertex** or an **edge** in a **graph**. It is obtained by counting the number of paths that pass through a **vertex** (site-betweenness) or through an **edge** (edge-betweenness) to connect every possible pair of vertices in the **graph**.
- **Bi-partite graph.** A **graph** is bipartite if there is a division (partition) of the vertices V into sets V_1, V_2 such that $V = V_1 \cup V_2$ and the two sets are independent (no vertices in V_1 are connected each other, and no vertices in V_2 are connected each other). In other words, I can colour the vertices in red and yellow such that no link exists between vertices of the same colour.
- **Centrality.** Also defined as the importance of a **vertex** or an **edge**. The **betweenness** is a measure of centrality. Another measure of centrality is obtained by considering the average distance of one **vertex** (or **edge**) from all the others.
- **Clustering.** A measure of edges density for a subgraph. The larger the clustering, the larger the density of edges with respect to the rest of the graph. Often the clustering reveals the existence of a **community**.
- **Clustering Coefficient.** The clustering coefficient of a **vertex** is given by the number of 'triangles' (**cycles** of length 3) it belongs to in the graph divided by the maximum possible number of triangles one could draw.
- **Community.** A set of vertices sharing similar properties. As an example, in the **WWW** people sharing similar interests tend to connect to the same HTML documents and tend to link to each other. In the overwhelming majority of cases a community results in large **clustering**.

- **Connectivity.** A graph is connected if a **path** exists between any two vertices in the graph. Sometimes it is erroneously used by some authors as equivalent to **degree**.
- **Cycle.** A closed **path** where the end-vertices coincide.
- **Degree.** The degree of a **vertex** is the number of its **edges**.
- **Diameter.** The maximum value of **distances** we can find in a **graph**.
- **Directed Graph.** a **graph** where edges have an orientation. Contrary to **oriented graph** multiple edges and **loops** are allowed.
- **Distance.** The distance of two **vertices** is given by the number of **edges** in a shortest path connecting them.
- **Edge, edges.** The connections between the **vertices** of the graph. They can be simple, multiple, oriented and weighted.
- **Food web.** The network composed of the species living in the same area connected by the predation relationships between them. In principle it is a directed graph where **loops** (cannibalism) and double **edges** (mutual predation) are allowed
- **Forest.** A set of disconnected **trees**
- **Fractal.** A **self-similar** geometrical figure. A geometrical structure is fractal when a portion has the same 'shape' than the whole. Also indicated as **self-similar**, that is 'similar' (in a mathematical sense) to itself. From the Latin *fractus* = broken, irregular.
- **Graph.** A mathematical object composed of **vertices** (sites) connected by **edges** (links).
- **HTML.** After 'hypertext mark-up language'. It is a way to write and link each other different documents. Most of the **WWW** pages are written in this way.
- **Internet.** The **network** of computers and routers linked each other by physical connections as cables or wireless connections. Not to be confused with the **World Wide Web**.
- **Leaf, leaves.** A vertex of degree one other than the **root** in a **tree**.
- **Loop.** An edge connecting one vertex with itself. Sometimes it is erroneously used instead of **cycle**.
- **Motifs.** All the possible **graphs** of a given 'little' (e.g. $3, 4, 5$ order. Their presence in larger networks can characterize the physical system described by the graph.
- **Multigraph.** A **graph** where it is allowed to have multiple **edges**.
- **Network.** In this book, any real system that can be completely described by a **graph**. In some cases the term **web** (**World Wide Web** or **food web**) it is used with the same meaning.
- **Order.** The number of vertices in a graph.

- **Oriented Graph.** A **graph** where edges have an orientation and contrarily to **directed graphs** multiple edges and loops are not allowed.
- **Path.** A series of consecutive edges in a graph.
- **Percolation.** A model for fractal growth in statistical physics. One medium is modelled by a regular grid. Sites of this grid belong to percolation clusters with probability p. When p is zero, there is no percolation cluster, when p is one, the entire medium belongs to percolation cluster. In between there is a value p_c above which a percolation cluster spans the medium. At the threshold value the cluster is **fractal**.
- **Probability density.** When the possible outcomes x of a measurement are real numbers, the probability density $P(x)$ gives the probability to obtain a value between x and $x + dx$.
- **Probability distribution.** It is the set of the probabilities $P(k)$ for the various outcomes k in a discrete space of events.
- **River network.** The graph obtained by considering the paths followed by water from rainfall on a specific area.
- **Root.** One special vertex in a tree (for example it can be the outlet of a **river network**. In this case the root is the only vertex with out-degree zero).
- **Scale-invariance.** A property of a **fractal** or **self-similar** object that does not change when the scale of observation is varied.
- **Scaling.** The behaviour of physical quantities when the size L of the system or the number of constituents N tends to infinite.
- **Self-affine.** An object with different **scaling** for different directions.
- **Self-similar.** An object that has the property of **scale-invariance**. Self-similar object have the same **scaling** for different directions. In this book the same as **fractal**.
- **Simple Graph.** A **graph** with neither multiple nor oriented edges.
- **Size.** The number of edges in a graph.
- **Taxonomy.** The classification of objects in the form of a hierarchical tree by successive steps of grouping.
- **Tree.** A **graph** without **cycles**. When spanning, it is a subgraph that connects all the *vertices* of the **graph**
- **Vertex, vertices.** The basic constituents of a **graph** together with the **edges**. Vertices represent the objects, individuals or species whose relationships are modelled through the **edges**. Those two sets of elements form a **graph**
- **World Wide Web.** The **network** of documents written mostly in **HTML** language, connected by hyperlinks. Not to be confused with the **Internet**.

Appendix B. Graph quantities

As already mentioned in the preface and introduction, this book is intended to be a guide in the area of networks for a large audience of readers. Our hope is that, people whose background is in economics or in biology will find this book useful. Therefore the mathematical description of graphs has been severely reduced compared with a standard text on graphs. Here we do not intend to provide all the notions of a mathematical book (partly because of the limited size of this appendix) but rather we want to give a more comprehensive description of the notions at the heart of graph theory. To make this appendix readable, we decided to write it as a self-contained structure. The price to pay is that some of the notions already presented in the text are repeated here. With all the limitations of the present space we follow here the exposition of Diestel (2005) and Bollobás (1979). These two texts should really be consulted by readers who wish to gain a more detailed knowledge of this branch of mathematics. Given this particular topic, this whole appendix must be regarded as a technical one (☢).

B.1 Basics

- A *graph* is a way to code a relation between the elements of two sets. These two sets are called V (set of *vertices*) and E (set of *edges*). The graph indicated as $G(V, E)$ can be drawn plotting the vertices as points and the edges as lines between them. It is not important how they are actually drawn. Ultimately the only thing that matters is to know which vertices are connected. For the moment we allow only one edge between a given pair of vertices.

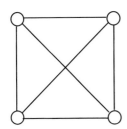
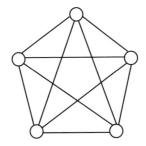

Fig. B.1 Two different examples of a complete graph. On the left K_4 and on the right K_5.

- A graph $G(V, E)$ where V has n elements (n vertices) is said to have *order* n. Analogously, the *size* of a graph is the number m of its edges (the number of elements of the set E).[79]
- When an edge e links vertices v_1, v_2 we have that vertices v_1, v_2 are *incident* with the edge e. Alternatively the edge e *joins* v_1, v_2 that are its *end vertices*.
- Vertices v_1, v_2 joined by edge e are *adjacent* or neighbours.
- A *dominating set* for a graph is a set of vertices whose neighbours, along with themselves, constitute all the vertices in the graph.
- A graph defined above whose *order* is n cannot have more than m_{max} edges where $m_{max} = n(n-1)/2$ (the *size* is smaller than m_{max}).[80] When all these possible edges are present the graph is *complete* and it is indicated with the symbol K_n (see Fig. B.1). The opposite case happens when there are no edges at all. The graph is then *empty* and it is indicated by the symbol E_n.

B.2   Different kinds of graphs

B.2.1   Weighted, directed, and oriented graphs

- Whenever a real number can be attached to an existing edge we have that the edge is characterized by a weight w. Note that in this book the weights will be almost exclusively positive real numbers, i.e. $w > 0$

---

[79] Since generalizations of this concept are possible and used, this class of graphs are sometimes called 'simple graphs'.

[80] The situation is different for directed graphs or multigraphs that we define in the next section.

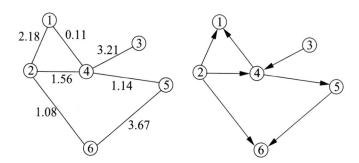

Fig. B.2 Left: a realization of a weighted graph. The degree of vertex 4 is 6.02 (given by $0.11 + 3.21 + 1.14 + 1.56$). Right: an example of an oriented graph. If the weight were the same in this case we would have an in-degree $k^{w,in} = 4.77(3.21 + 1.56)$ and an out-degree $k^{w,out} = 1.25(0.11 + 1.14)$.

(see left-hand side of Fig. B.2). The graph in this case is a **weighted** graph, shown A **directed** graph $G(V,E)$ is given by two disjoint sets E and V plus two functions $I(E \to V)$ and $F(E \to V)$. The first one assigns to every edge e an initial vertex $I(e)$. The second one assigns to every edge e a final vertex $F(e)$. More simply, every edge e has assigned a direction from one vertex $I(e)$ to another $F(e)$.

- Sometimes $I(e)$ and $F(e)$ coincide. In this case e is a *loop*. That is a loop is an edge between one vertex and itself. Moreover, we can have different edges directed between the same two vertices $I(e)$ and $F(e)$. This is the case of *multiple edges*.

- Whenever the direction is assigned but neither loops nor multiple edges are present, then the graph is **oriented**. Intuitively oriented graphs are undirected graphs where for every edge one assigns a direction.

- A **multigraph** is a pair of disjoint sets (V,E) together with a map $E \to V \cup [V]^2$ assigning to every edge either one or two vertices (the ends). A multigraph is then similar to a directed graph, with multiple edges and loops but no direction assigned. A sketch of the various kind of graph is presented in Fig. B.3.

- The number of edges of vertex v_i in a graph is called the **degree** of vertex v_i and is indicated here by $k(v_i)$. In the case of an oriented graph the degree can be distinguished in *in-degree* $k^{in}(v_i)$ and the *out-degree* $k^{out}(v_i)$. In the case of weighted graphs we will consider the *weighted-degree* $k^w(v_i)$ of a vertex v_i as the sum of the weight of the edges on v_i.

- If the set V in graph $G(V,E)$ is composed of vertices $v_1, v_2, ..., v_n$ then the series $k(v_1), k(v_2), ..., k(v_n)$ is a *degree sequence* of $G(V,E)$. Particular importance in this book is devoted to the statistical properties of

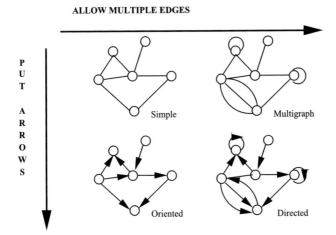

Fig. B.3 The distinction between the various types of graphs.

such degree sequence. In most real networks the frequency $P(k)$ with which a particular value k of $k(v_i)$ is present in the sequence follows a power law of the kind $P(k) \propto k^{-\gamma}$. This case is indicated by scale-free graph.

B.2.2   Subgraphs

- Consider two graphs $G(V,E)$ and $G'(V'E')$. We can define a new graph indicated by $G \cap G'$ whose vertices are in the set $V \cap V'$ and the edges in the set $E \cap E'$. These operations are shown in Fig. B.4. If $V \cap V' = \emptyset$ the two graphs are *disjoint*. On the other hand if $V' \subseteq V$ and $E' \subseteq E$ then $G'(V', E')$ is an *induced subgraph* of $G(V, E)$ and we indicate this by writing $G'(V', E') \subseteq G(V, E)$. Finally, if $G'(V', E') \subseteq G(V, E)$ and $V' = V$, $G'(V', E')$ is *spanning* of $G(V, E)$
- Two graphs are *isomorphic* if you can redraw one of them so that it looks exactly like the other.

B.2.3   Partitions in graphs

- Let $r \geq 2$ be an integer, a graph $G(V, E)$ is called *r-partite* if it can be divided into r classes such that every edge has its ends in different classes. This means that vertices in the same class cannot be adjacent. If $r = 2$ the graph is also called **bipartite**
- A complete *bipartite clique* $K_{i,j}$ is a graph where every one of i nodes has an edge directed to each of the j nodes.

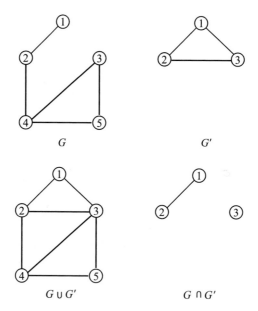

Fig. B.4 The operations of union and intersection of two graphs.

- A *bipartite core* $C_{i,j}$ is a graph on $i+j$ nodes that contains at least one $K_{i,j}$ as a subgraph.

B.3   Paths, cycles, and trees

- A **path** is a non-empty graph $G'(V', E')$ of the form $V' = v_0, v_1, ..., v_n,$ $E' = e_1,, ...e_n$ where $v_0, v_1, ..., v_n$ a set of vertices for which e_i is an edge joining vertices v_{i-1} and v_i. Less formally we can say that a series of consecutive edges forms a *path*. The number of edges in a path is called the *length* of the path.
- if $P = e_1 + e_2 + ... + e_n$ is a path then if $n \geq 3$ and we add an edge e_0 joining vertices v_n and v_0, we obtain a *circuit*. In other words a *circuit* is a path whose end vertices coincide. If in the circuit all the vertices are distinct from each other the circuit is a *cycle*. A cycle of length k is indicated as C^k. Note that *a cycle is different from a loop*.

  - A *Hamiltonian path* shown in Fig. B.5 left, is a path that passes once through all the vertices (not necessarily through all the edges) in the graph. A Hamiltonian circuit is a Hamiltonian path that begins and ends in the same vertex. By construction this circuit is also a cycle.

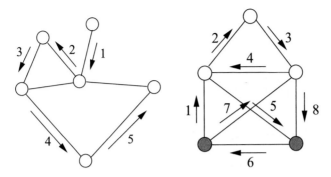

Fig. B.5 Left: a Hamiltonian path. Right: an Eulerian circuit.

- An *Eulerian path*, shown in Fig. B.5 right, is a path that passes once through all the edges (not necessarily once through all the vertices) in the graph. An Eulerian circuit is an Eulerian path that begins and ends in the same edge. If the vertices in the circuit are all different then the circuit is a cycle.

- When a path exists between any couple of vertices v_i, v_j in a graph, the graph is *connected*. This property called *connectivity is often misused by some authors in the sense of degree*. As shown the two quantities are very different.

B.3.1   Trees

- A *tree* is a connected graph that does not contain cycles (also known as acyclic graph). If the graph is not connected but still acyclic then it is composed of different trees and assumes the natural name of *forest*.
- Vertices of degree 1 in a tree are called *leaves*. In any non-trivial tree there are at least two leaves.
- Sometimes it is convenient to consider one vertex of the tree as special. This vertex is called the *root*. A tree with a fixed root is a *rooted tree*.

Appendix C. Basic statistics

Statistics deals with chance. This appendix does not want to (nor could it), provide a course in this field. Rather it is a collection of some of the basic notions used in the book. Readers not familiar with statistics are strongly suggested to consult specific textbooks. This appendix is therefore only aimed at those readers who are familiar with some of the concepts presented. Because of this approach we will not enter into the great mathematical difficulties associated with the foundations of probability. For our purposes we call the probability of an event the limit (as N goes to infinity) of the ratio between the number of appearances of this event and the total number of observations N.

C.1   Events and probability

We indicate an *event* as one of the possible states of a varying physical system. The paramount example used in all textbooks is that of dice. The roll of one die is the varying system, and the possible states are the values of the six faces. If the event i happened N_i times in N trials we indicate as $P(i) = \lim_{N \to \infty} N_i/N$ the *probability* of the event i. In other words the probability $P(i)$ is the limit of the ratio N_i/N when the number of trials goes to infinite. If the outcome of the event is the result of a die thrown, then all the possible results are $1, 2, 3, 4, 5,$ and 6. Since one of the six must happen, whatever is the series of the observations

$$P(1) + P(2) + P(3) + P(4) + P(5) + P(6) = \sum_{i=1,6} P(i) = 1 \qquad \text{(C.269)}$$

must hold. This equation simply states that states 1 to 6 fully complete the series of possible results. The set of natural numbers $1, 2, 3, 4, 5, 6$ is called the space of events. In principle every face in a fair die is equally likely. This consideration can be written in formulas as

$$P(1) = P(2) = P(3) = P(4) = P(5) = P(6). \qquad (C.270)$$

This equation coupled with the previous one, means that $P(i)$ for every i must be $1/6$. In general no die is perfect and if we put on a histogram the values of the 'real' $P(i)$ in an experiment we can discover that the slope is not perfectly flat. The previous histogram gives the form of the function $P(i)$. This quantity is called the probability function or **probability distribution** and it is a fundamental tool for characterizing the various chances. A flat function is a uniform probability distribution and in the case of dice gives some indication on the fairness of the process. We see in the following that in different situations those shapes can be different from the flat one. For example, if event i is never recorded then $P(i)$ is equal to 0. If every trial gives the event i then the probability has its maximum value and $P(i) = 1$.

C.2   Probability densities and distributions

Sometimes the space of events is not a discrete series of natural numbers, but instead corresponds to a portion of the real axis. If we measure the height of an individual in a population we might obtain for example 1.78 metres. Maybe if our measuring device is precise enough we can obtain a number like 1.7852 metres or even more precise (the only limitation is given by the precision of the measurement).

Now the situation is rather different from the previous case. Rolling a dice we can obtain 1 or 2, but we cannot have as a result 1.5. Measuring heights we can have in principle any real number. The histogram of the probability values becomes a function of a real variable x. Actually is a lot simpler to measure the probability of having a height between 1.77 and 1.78 metres (it turns out that having an infinitely precise ruler the probability of measuring *exactly* 1.7731415927 metres or any other real number goes to 0 as the measure precision increases). In this case we use the **probability density function** $f(x)$. When x is a continuous random variable, then the probability density function is a function $P(x)$ such that for two numbers, a and b with $a < b$, the probability that the event will be comprised between a and b is a number given by

$$P(a < x < b) = \int_a^b f(x)dx. \qquad (C.271)$$

It is important to note that while the probability distribution $P(k)$ has the meaning of a probability, the probability density $f(x)$ does not. If we have $P(k) = 0.1$, this means that on average 10 per cent of the times we find a degree equal to k. In the case of the probability density $f(x)$ the probability of extracting exactly a precise real number (x) is always zero. The correct approach is to define a quantity that gives the probability that the variable x is within an interval dx around the expected value. The quantity that gives the probability is now $f(x)dx$. The name 'density' refers to the fact that one recovers the meaning of probability only by multiplying it by a suitable interval. One of the commonest $F(x)$ one encounters in nature is the bell-like shape of the Gaussian (normal) probability density function .

$$f(x) = \frac{1}{\sqrt{2\pi\sigma^2}} e^{\frac{(x-\mu)^2}{2\sigma^2}} . \tag{C.272}$$

Here we have an expected value for x given by μ. The deviation from this mean value is given by the σ. This means that we expect with a 68.3 per cent confidence that any x will be between $\mu - \sigma$ and $\mu + \sigma$. In 95.4 per cent of the cases x will be between $\mu - 2\sigma$ and $\mu + 2\sigma$ and finally in 99.7 per cent of the cases the value will be between $\mu - 3\sigma$ and $\mu + 3\sigma$.

C.2.1   The example of the graphs

Simple graph quantities refer to the properties of a single vertex (or in the case of motifs to a small group of vertices). Since many networks of interest are composed of thousands of vertices, the above information for all the elements would be practically intractable. That is the reason that forces, to give a statistical description of the system. Focusing on the degree, for example, we can consider the average of the various degrees and study its mean value. Much more information is stored in the frequency distribution of the various values of the degrees. This means that we compute how many times (with respect to the total number of vertices N) we find one vertex with degree $1, 2, 3$, etc. It is important to note (it has already been mentioned in the introduction) that the only data available for many networks are the frequencies of the various degrees. The limit value after infinite measurements of the proportion of the various frequencies tends to their probabilities.

Before proceeding further there is another important thing to discuss. We should in principle consider the properties of a discrete series of values $P(1), P(2), P(3), \dots$ etc. (giving the probability to find a vertex with degree $1, 2, 3, \dots$ respectively). It is instead customary to use a continuous function $f(k)$ treating k as a real number. This means that $f(k)$ is also defined for non-integer values of its argument k. In this book we follow this approximation because it helps in obtaining some analytical results. An example of this approximation is represented in Fig. C.1 where the histogram of the various discrete probabilities is fitted through a suitable continuous function.

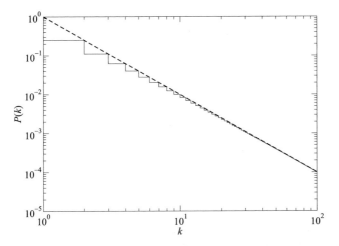

Fig. C.1 The histogram of a discrete probability function and its approximation through a continuous function. Since we have unitary steps, the relative error is $\frac{dk}{k} = \frac{1}{k}$, which in the limit of large k tends to zero.

In the large degree limit the variation dk (actually ± 1) can be considered negligible.[81] There is an immediate advantage in this approach that consists in the possibility of describing the histogram of probability distribution by means of a more compact formula giving the form of the probability density. Whenever in this book we say that the probability distribution $P(k)$ varies as $k^{-\gamma}$ we intend exactly this approximation by means of continuous functions.

Through probability density it is also easier to realize the integration of the function $f(k)$. The integral of the $f(x)$ that we indicate as $f^>(x)$ gives the probability that a certain event has an associate probability greater than or equal to x. Since, in most cases, data are rather noisy, fluctuations above or below the expected trend can be very strong. Integration should then average away these fluctuations and allows us to remove the noise. In the case of ubiquitous (at least in this book) power-law densities the integration is particularly simple provided we have $\gamma > 1$, and $x_0 > 0$.

$$f(x) = Ax^{-\gamma} \rightarrow f^>(x) = A \int_{x_0}^{\infty} x^{-\gamma} dx = Ax^{-\gamma+1}. \qquad (\text{C.273})$$

---

[81] That is the case of huge graphs as the WWW.

It has to be noticed though that samples are sometimes very small. It is therefore unlikely to deal with infinity as upper cut-off to a good approximation. In those cases, analytically, the form of the $f^>(x)$ is rather given by

$$f^>(x) = Ax^{-\gamma+1} - AC^{-\gamma+1}, \qquad (C.274)$$

where C is the value of the real cut-off. For large values of C we found of course the limit case, in the other case this adding term is responsible for a noticeable bending of the function (see Section 3.4.2 and particularly eqn 3.89).

C.2.2   Playing with probability densities

In this book we very often use power-law distributions. For some applications it can be useful to produce such distributions with the computer. The most common random number generators produce random numbers with a uniform distribution. This means that the probability of extracting a number between 0 and 0.5 is the same as extracting a number between 0.5 and 1. A little trick is necessary to transform these uniformly distributed numbers into values following a power-law shape. The main idea is that outcomes of experiments do not really depend upon the variables we use to describe them. If we have a variable x representing the outcome of an experiment (e.g. the number on a roulette wheel) we can decide to study this phenomenon by means of any variable $y = g(x)$. As a result the probability of the event must not change.

Let us consider a simple example: the number of times a ball stops in a certain value x is given by $f_x(x)dx$. If we want to use the variable $y = g(x)$ to describe the same event then $f_y(y)dy$ must return the same value (where f_y is a function different from f_x). This determines the form of the distribution of the function $f_y(y)$. In the limit of $dx \to 0$ and $dy \to 0$ we have

$$f_x(x)dx = f_y(y)dy \to f_y(y) = \frac{f_x(x)}{dy/dx} = \frac{f_x(x)}{|g'(x)|} \qquad (C.275)$$

where $g'(x)$ is the derivative of $g(x)$ with respect to x.

In the above example we can check how to pass from uniform ($f_x(x)$ is constant) to power-law distribution ($f_y(y) = y^\beta$). It is easy to check that a change of variable $y = g(x) = x^\gamma$ produces what we are looking for

$$f_y(y) = \frac{f_x(x)}{\gamma x^{\gamma-1}} = Ax^{1-\gamma} = Ay^{(1-\gamma)/\gamma} = Ay^\beta. \qquad (C.276)$$

where $\beta = (1-\gamma)/\gamma$ and A is a constant given by $f_x(x)/\gamma$. Therefore, if we want a $f_y(y)$ with a specific value of β we have to raise to the power

$\gamma = 1/(1 + \beta)$ a random number x extracted from the commonly used uniform distribution.

C.3   Working with statistical distributions

Coming now back to the plot in Fig. C.1 we see that the axes are in logarithmic scale. The reason for such a choice of scale is related to the topic of this book. In almost all cases of interest the histogram of the discrete probabilities $P(k)$ and its best-fitting continuous function $f(x)$ has the shape a power law. This corresponds to saying that for the discrete distribution $P(k)$ we have

$$P(k) \propto k^{-\gamma}. \qquad (C.277)$$

The symbol $\propto$ means 'proportional to' and this relational concept is used very often in this book.

From now on we no longer distinguish between the $P(k)$ and its corresponding real function $f(k)$, this corresponds to considering from now on the degree as a continuous variable. Correspondingly, we shall indicate both as $P(k)$.

Probability density and probability distributions obey the normalization condition. In other words we are sure (probability $=1$) that we will find at least one value of the degree whatsoever.

☛ In this specific case this means that the frequency density is in fact $P(x) = Ax^{-\gamma}$. The proportionality constant A is however not important. It is fixed by the requirement that the sum of all the frequencies is equal to one. This means

$$\int P(x)dx = \int Ax^{-\gamma}dx = 1 \qquad (C.278)$$

$$\rightarrow A = \frac{1}{\int x^{-\gamma}dx}. \qquad (C.279)$$

Some care must be used with the extremes of integration. While the upper limit can be considered infinite, in the lower limit if $x = 0$ the integrand diverges. Therefore if the $F(x)$ is the probability density for the degree x (treated as a real variable), when computing A we have to remember to restrict ourselves to the connected part of the system, that is where $x \geq 1$. Whenever extremes of integration are not explicitly indicated we assume the above conditions apply.

The knowledge of the distribution function is particularly important. Through this quantity we can compute a couple of other interesting quantities. For example we can compute what is the typical ('mean') value that the degree assumes in the graph. This value will be indicated by $\langle x \rangle$, where the symbol $\langle ... \rangle$ indicates an average over all the possible outcomes.

A measure of the typical error we make if we assume that every vertex has degree $\langle x \rangle$ (thereby neglecting values fluctuations in our system) is given by the variance σ^2.

☕ These two quantities are given by definition by

$$\langle x \rangle = \int x P(x) dx \tag{C.280}$$

$$\sigma^2 = \int (x - \langle x \rangle)^2 P(x) dx \tag{C.281}$$

C.3.1   Some important examples

It is easy to see that if the integral of the $P(x)$ must be equal to one and x can vary between 1 and infinity, at a certain point the function $P(x)$ must decrease fastly, otherwise its integral would diverge.

Amongst all the possible functions obeying such requirements the real distribution probabilities are remarkably few. Here we will further restrict this choice to the ones that are more frequent.

- A discrete probability distribution is the Poisson distribution $P_P(k)$ (where k is integer) behaving as

$$P_P(k) = \frac{\mu^k e^{-\mu}}{k!}. \tag{C.282}$$

- If the variable k of the function $P(k)$ also assumes real values, we have a series of different possible probability densities (remember that we now use the same symbol $P(k)$ used for the probability distributions). For our purpose it is enough to mention:

    The Gaussian (or normal) density function $F_G(x)$

$$P_G(k) = \frac{1}{\sqrt{2\pi}\sigma} e^{-\frac{(k-\mu)^2}{2\sigma^2}}. \tag{C.283}$$

    The log-normal density

$$P_{LN}(k) = \frac{1}{\sqrt{2\pi}\sigma k} e^{-\frac{(\ln(k)-\mu)^2}{2\sigma^2}}. \tag{C.284}$$

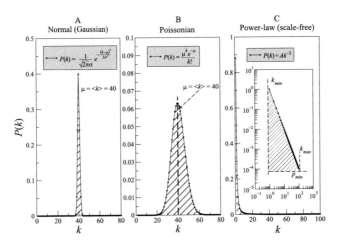

Fig. C.2 (A) The plot of a Gaussian Distribution. (B) The plot of a Poisson distribution. (C) The plot of a power-law distribution. In the inset of (C) the same plot on a logarithmic scale. Note that the power-law function is not defined in $k = 0$.

The power-law density $P_{pl}(k)$

$$P_{pl}(k) = Ak^{-\gamma}. \tag{C.285}$$

The Gaussian distribution has mean μ and variance σ^2, the log-normal instead is more skewed with mean $e^{\mu+\sigma^2/2}$ and variance $e^{2\mu+\sigma^2}(e^{\sigma^2} - 1)$. The power-law distribution is the only one that may have no finite mean and variance.

We see in Fig. C.2 a snapshot of the normal, the Poisson, and the power-law distributions. Note that the first two functions increase to a maximum after which they decay. The third has a smoother character always decreasing as k grows. Note also that normal and Poisson distribution depend upon the choice of some parameter, even if their qualitative behaviour does not change. Note that the dashed area in the three plots of Fig. C.2 has a value of 1. That is another way of saying that the three distributions are normalized.

Interestingly (we return to this point), while it is easy to distinguish between a Gaussian and a power-law distribution, it can be difficult to spot the difference between a power-law function and a log-normal function. Most of the time even eye inspection can help in determining which function is most suitable for the data. A power-law distribution will look like a straight line whenever plotted on a logarithmic scale (i.e. whenever considering the logarithms of both sides). Unfortunately, in some conditions this can be true also for the log-normal distribution.

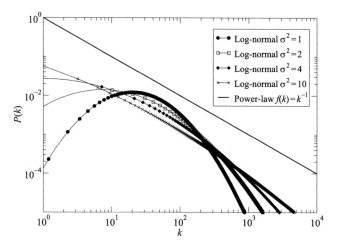

Fig. C.3 A comparison between log-normal distributions with different variances and a power law whose exponent is -1 (in this example the average μ is equal to 4, so that the very first part should decay as $k^{-0.6}$. Note that the range of k is typical of many data sets available. Therefore the true tail of a log-normal distribution may not be noticeable in some cases. Note also that the change of gradient is so slow that even the intermediate region recalls the behaviour of a power law of larger slope (about -1).

 Taking the natural logarithm of eqn C.285 we have

$$\ln(P_{pl}(k)) = \ln(A) - \gamma \ln(k); \qquad (C.286)$$

plotting $\ln(P_{pl}(k))$ vs $\ln(k)$ we have a linear relationship with slope $-\gamma$. Now making the same for eqn C.284 we have

$$\ln(P_{LN}(k)) = -\ln(\sqrt{2\pi}\sigma) - \ln(k) - \frac{(\ln(k) - \mu)^2}{2\sigma^2} \qquad (C.287)$$

$$= -(\frac{\mu^2}{2\sigma^2} + \ln(\sqrt{2\pi}\sigma)) + (\frac{\mu}{\sigma^2} - 1)\ln(k) - \frac{\ln(k)^2}{2\sigma^2}$$

where the term $-(\frac{\mu^2}{2\sigma^2} + \ln(2\pi\sigma))$ has the same role as the $\ln(A)$ in eqn C.286 representing the constant term.

Whenever the log-normal distribution is characterized by a value $\sigma \gg \ln(k)$ then the term $\frac{\ln(k)^2}{2\sigma^2}$ in eqn C.288 vanishes and the behaviour becomes very similar to that of a power-law function as shown in Fig. C.3. Note that this feature does not change by changing the base of the logarithm. This only introduces a constant term on both sides of the equations. The *apparent*

slope is then given by

$$\gamma_{app} = (\frac{\mu}{\sigma^2} - 1). \tag{C.288}$$

Of course as we go along the tail, the logarithm of k grows and the i quadratic term becomes less and less negligible. Nevertheless, since the logarithm is a slowly growing function, one may need a very large interval on the axes, in order to clearly distinguish between the two functions.

This is an important point, and it is worth discussing it for a moment. We are particularly interested in the power law because of its **scale-invariant** or **scale-free** properties. This means that if we change the scale of observation (i.e. if we consider a change of scale of the kind $k' = ak$) the only distribution that maintains its analytic form unaltered is the power law.

C.3.2   Finite sizes

Recalling the discrete nature of these functions and their physical meaning we would like to stress here that even scale-free functions (see inset of Fig. C.2 C) have a particular scale when considering real data. This is given by the finite size of our samples. In the specific case of networks we must have a minimum degree of one (you cannot have less than one edge) and a maximum degree that is the order of the network (you cannot have more than $n - 1$ edges). Hoping that (as in the case of the WWW) the graph is very large we can shift the upper cut-off as far as we want along the x-axis. Still, we must expect a deviation from power-law behaviour whenever k is similar to the size of the system.

In the case of a power-law form of the $P(k)$ we obtain for the normalization condition

$$\langle k \rangle = A \int kk^{-\gamma} dk = \frac{\int k^{1-\gamma} dk}{\int k^{-\gamma} dk}. \tag{C.289}$$

Since for the real world we never find an exactly infinite degree, we must stop integration at a large value k_{max} (k_{max} of the order of N). The above equation then becomes

$$\langle k \rangle = \frac{\int_1^{k_{max}} k^{1-\gamma} dk}{\int_1^{k_{max}} k^{-\gamma} dk}, \tag{C.290}$$

which is finite (i.e. does not diverge as k_{max} grows) for $\gamma > 2$.

 In order for the $P(k)$ to be a probability function it must be integrable, or in other words the A must be finite. This gives a first condition $\gamma > 1$. We see now that there are two possibilities. Let us consider the result of eqn C.290, that is

$$\langle k \rangle = \frac{\int_1^{k_{max}} k^{1-\gamma} dk}{\int_1^{k_{max}} k^{-\gamma} dk} = \frac{(1-\gamma)}{(2-\gamma)} \frac{k_{max}^{2-\gamma} - 1}{k_{max}^{1-\gamma} - 1}. \tag{C.291}$$

- $\gamma < 2$ In this case the above expression eqn (C.290) grows as $k_{max}^{2-\gamma} - 1$.
- $\gamma > 2$ In this case instead the above expression as k_{max} increases, tends to the constant value $\frac{1-\gamma}{2-\gamma}$. The average $\langle k \rangle$ stays finite no matter how large is the cutoff.

An analogous behaviour holds also for the variance. This quantity remains finite whenever the exponent γ is larger than 3. Instead it diverges as the cut-off k_{max} increases.

This can be seen with a derivation similar to the one already made for the mean. In this case we find

$$\sigma^2 = \int_1^{k_{max}} (k - \langle k \rangle)^2 P(k) dk = A \int_1^{k_{max}} k^{2-\gamma} dk - \langle k \rangle^2. \tag{C.292}$$

Therefore the above equation has the solution

$$\sigma^2 = \frac{(1-\gamma)}{(3-\gamma)} \left(\frac{k_{max}^{3-\gamma} - 1}{k_{max}^{\gamma-1} - 1} \right) - \langle k \rangle^2. \tag{C.293}$$

Again we have that in the case of $\gamma < 3$ the variance grows as

$$\sigma^2 = \frac{(1-\gamma)}{(3-\gamma)} \left(k_{max}^{3-\gamma} - 1 \right) - \langle k \rangle^2. \tag{C.294}$$

Instead for $\gamma > 3$ the variance tends to a constant value regardless of the value of the cut-off k_{max}. In all these cases it makes little sense to describe the system by means of average values.

We see in the following that almost every scale-free network is characterized by a power-law degree distribution with a value of γ between 2 and 3. This means that even if it were possible to define a finite average, the standard error for this value would be of the order of magnitude of the size of the system.

C.4   Statistical properties of weighted networks

As we saw in the second part of the book, in most real situations the networks that one finds are weighted (Yook et al., 2001; Garlaschelli et al., 2005) .

While the generalization of the concept is straightforward, we have already seen that sometimes it is not very easy to generalize the definition of the quantities, as in the case of the clustering coefficient. Whenever these quantities are defined, anyway, we are interested in their statistical distributions. Also in this case the weighted degree is usually power-law distributed. This is what we have already found in the 'simple' network analysis, interestingly, in the case of weighted networks new scale-free relations arise.

This means that if we consider the degree k_i^w for a vertex i, where

$$k_i^w = \sum_{j=1}^{n} a_{ij}^w, \qquad \text{(C.295)}$$

the $P(k^w)$ is power-law distributed. Incidentally note that since the value of k^w is a real number, we have a continuous probability function $P(k^w)$.

Also for the strength density function, we find a behaviour similar to that of the degree. This is not surprising because we have a connection between the two quantities given by the scaling relation

$$s(k) \propto k^{\zeta} \qquad \text{(C.296)}$$

where now $s(k)$ is the average strength for vertices whose degree is k. In the case of absence of correlation between the two quantities one is allowed to take the averages on both sides of eqn C.295 finding

$$s(k) = \frac{1}{2m} \sum_{i,j=1}^{N} a_{ij}^w k = \langle w \rangle k \qquad \text{(C.297)}$$

where m is the total number of edges in the network and $\langle w \rangle$ is its average weight. In this case the value of ζ is equal to one. It can be found (Barrat et al., 2004a) a value of ζ between 1 and 1.5.

A more general approach is to consider a weighted matrix as a matrix of probabilities (Ahnert et al., 2006). These probabilities form an *ensemble of edges*, or more concisely, an *ensemble network*. Thus, just as any binary square matrix can be understood as an unweighted network and any real square matrix corresponds to a weighted network, any square matrix with entries between 0 and 1 can be viewed as an ensemble network. If we sample each edge of the ensemble network exactly once, we obtain an unweighted network which we term a *realization* of the ensemble network. In particular, p_{ij} is the probability that the edge between nodes i and j exists, where

$$p_{ij} = \frac{w_{ij} - \min(w_{ij})}{\max(w_{ij}) - \min(w_{ij})}. \qquad \text{(C.298)}$$

In an ensemble network, the corresponding sum over the edges attached to a particular node gives the *average degree* of node i *across realizations*, denoted as $\langle k_i \rangle$ and given by $\langle k_i \rangle = \sum_j p_{ij}$.

It is important to note that while the strength of a node in a weighted network may have meaning in the context of the network, $\langle k_i \rangle$ has a universal meaning, regardless of the original meaning of the weights. Now consider the total number of edges n in a network (also referred to as its *size*) given by $n = \sum_{ij} a_{ij}$. Replacing a_{ij} by p_{ij} again gives us the average size $\langle n \rangle$ of the realizations of the ensemble network, which is simply $\langle n \rangle = \sum_{ij} p_{ij}$.

A more complex measure in unweighted networks is the average degree of the nearest neighbours k_i^{nn}, which is the number of neighbours of i's neighbours, divided by the number of neighbours of i:

$$k_i^{nn} = \frac{\sum_j k_j}{k_i} = \frac{\sum_{j,k} a_{ij} a_{jk}}{\sum_j a_{ij}} \tag{C.299}$$

where $j \neq i$ in the sums. By rewriting k_i^{nn} solely in terms of the a_{ij}, this generalizes to ensemble networks in a very straightforward manner:

$$k_i^{nn,e} = \frac{\sum_{j,k} p_{ij} p_{jk}}{\sum_j p_{ij}}. \tag{C.300}$$

This measure $k^{nn,e}$ is simply a ratio of averages: the average number of neighbours of i's neighbours over the average number of i's neighbours. For unweighted networks the *clustering coefficient* of a node i has been defined as:

$$c_i = \frac{\sum_{j,k} a_{ij} a_{jk} a_{ik}}{k(k-1)/2} = \frac{\sum_{j,k} a_{ij} a_{jk} a_{ik}}{\sum_{j,k} a_{ij} a_{ik}} \tag{C.301}$$

where $k \neq j \neq i \neq k$ in the sums. This corresponds to the number of triangles in the network which include node i, divided by the number of pairs of bonds including i, which represent 'potential' triangles. Using the ensemble approach with its normalized weights this generalizes straightforwardly to:

$$c_i^e = \frac{\sum_{j,k} p_{ij} p_{jk} p_{ik}}{\sum_{j,k} p_{ij} p_{ik}} \tag{C.302}$$

which can be read as the average number of triangles divided by the average number of bond pairs.

Appendix D.   Matrices and eigenvectors

We just recall briefly here that a system of linear equations of the type

$$\begin{cases} a_{11}x_1 + a_{12}x_2 + \dots + a_{1n}x_n & = & y_1 \\ \quad\quad\dots. \\ a_{n1}x_1 + a_{n2}x_2 + \dots + a_{nn}x_n & = & y_n \end{cases}$$

can be written in the compact form $\mathbf{Ax} = \mathbf{y}$ through the introduction of the matrix $\mathbf{A}$, and vectors $\mathbf{x}, \mathbf{y}$, where

$$\mathbf{A} = \begin{pmatrix} a_{11} & a_{12} & \dots & a_{1n} \\ a_{21} & a_{22} & \dots & a_{2n} \\ a_{31} & a_{32} & \dots & a_{3n} \\ \dots & \dots & \dots & \dots \\ a_{n1} & a_{n2} & \dots & a_{nn} \end{pmatrix} \qquad \text{(D.303)}$$

$$\mathbf{x} = \begin{pmatrix} x_1 \\ x_2 \\ x_3 \\ \dots \\ x_n \end{pmatrix} \qquad \text{(D.304)}$$

$$\mathbf{y} = \begin{pmatrix} y_1 \\ y_2 \\ y_3 \\ \dots \\ y_n \end{pmatrix} . \qquad \text{(D.305)}$$

The product $\mathbf{Ax}$ between the matrix $\mathbf{A}$ and the vector $\mathbf{x}$ is defined so as to recover the expression in eqn D.303. More particularly, we obtain one element for every row of $\mathbf{A}$, those elements are computed by summing the product of n-th element on the row, with the corresponding n-th element on the only column of $\mathbf{x}$. This operation is possible only when the number of columns of $\mathbf{A}$ is equal to the number of rows of $\mathbf{x}$. Of course the components of matrix $\mathbf{A}$ are exactly the $n \times n$ coefficients of the n variables in eqn D.303. Similarly the variables x are listed in what is called a vector (it can be considered a matrix of size $n \times 1$).[82] The basic operations of vectors and matrix manipulations we need in the book are listed below.

- By multiplying a matrix $\mathbf{A}$ with a real number λ we obtain a matrix whose elements are the original multiplied by λ.
- As already seen, by multiplying a matrix $\mathbf{A}$ of $n \times n$ elements with a vector $\mathbf{x}$ of $n \times 1$ elements we obtain another vector $\mathbf{y}$ of $n \times 1$ elements. The first element of vector $\mathbf{y}$ is the sum of the product of the elements on the first row of matrix $\mathbf{A}$ with the column $\mathbf{x}$. That is to say

$$y_1 = a_{11}x_1 + a_{12}x_2 + \dots + a_{1n}x_n \tag{D.306}$$

and so on for the other elements.
- By multiplying with matrices $\mathbf{A}, \mathbf{B}$ of $n \times n$ elements with each other we obtain another matrix $\mathbf{C}$ of the same number $n \times n$ of elements. The element c_{ij} on row i and column j is given by following the same rule described above, that is by multiplying each element of row i with every element of column j. That is,

$$c_{ij} = a_{i1}b_{1j} + a_{i2}b_{2j} + \dots + a_{1n}b_{nj}. \tag{D.307}$$

- An eigenvector of a matrix A is a vector $v = (v_1, v_2, \dots, v_n)$ such that $Av = \lambda v$, that is

$$\begin{pmatrix} a_{11} & a_{12} & \dots & a_{1n} \\ a_{21} & a_{22} & \dots & a_{2n} \\ a_{31} & a_{32} & \dots & a_{3n} \\ \dots & \dots & \dots & \dots \\ a_{n1} & a_{n2} & \dots & a_{nn} \end{pmatrix} \begin{pmatrix} v_1 \\ v_2 \\ v_3 \\ \dots \\ v_n \end{pmatrix} = \lambda_v \begin{pmatrix} v_1 \\ v_2 \\ v_3 \\ \dots \\ v_n \end{pmatrix} \tag{D.308}$$

where λ_v is the relative eigenvalue. In other words the eigenvectors of a certain matrix are not transformed by the application of the matrix. The only effect is that they are multiplied by a number λ. The eigenvalues λ are in general complex numbers, though in some special cases they can be real numbers. Finally, eigenvalues can be degenerate; that is, more than one eigenvector can be associated with them.

---

[82] We denote by convention the first entry as the number of rows and the second entry as the number of columns.

Appendix E. Population dynamics

E.1 Population dynamics

E.1.1 One species, resources infinite

This problem was first studied by the Italian mathematician Fibonacci in 1202. He made a series of different approximations in order to describe the following simple 'ecosystem'

- We start with a pair of rabbits.
- Time is divided into discrete steps ('months'). At the end of every 'month', the rabbits mate; at the end of the next 'month' every couple produces a new couple (male and female). The newborn are able to mate after only one month. In the steady state every couple produces a new couple per month, only the newborn wait one month to grow up.
- Birth rate is much larger than death rate. Therefore as a first approximation we do not take into account deaths. Furthermore rabbit numbers can grow indefinitely. In other words, the resources of the environment are infinite.

With the rules described above, it easy to understand that at the first month we have the first couple of rabbits growing up, at the second month they mate and finally at the third month there is a new couple. At the fourth month the old couple produces a new couple while the newborn grow up. In general the number of couples at month t is given by those at the previous time step (none dies) plus a number of newcomers given by couples grown up (all those present two time steps earlier). By introducing the quantity $N(t)$ that gives the number of couples at time t we have

$$N(t) = N(t-1) + N(t-2) \qquad \text{(E.309)}$$

This results in the world-famous and ubiquitous Fibonacci series $1, 1, 2, 3, 5, 8, 13, \ldots$ where every entry is given by the sum of the previous

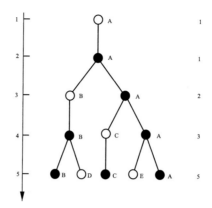

Fig. E.1 The tree of rabbit growth. The couples who can have offsprings are filled in black. Young couples are empty circles. Along the tree you can see the story of the original couple A (which does not die). Left: the number of generation. Right: the population of the species.

two as shown in Fig. E.1. In this approximation the death rate d is equal to zero. Therefore the growth rate g defined as the difference between the birth rate b and death rate d in this case is given only by the birth rate.

 By considering the number $N(t)$ of couples at time t, it is possible to compute the growth rate g. This is also defined as the difference of the population at two time steps, that is

$$g \equiv \frac{N(t+1) - N(t)}{N(t)} = \frac{N(t+1)}{N(t)} - 1. \tag{E.310}$$

Since in the specific case of the Fibonacci series $N(t+1) - N(t)$ is equal to $N(t-1)$, this means that by using eqn E.310 we can write

$$\frac{N(t+1)}{N(t)} - 1 = \frac{N(t-1)}{N(t)}. \tag{E.311}$$

If we assume that in the limit $t \to \infty$ the growth rate g exists, it must obey eqn E.311. Therefore:

$$g = \frac{1}{g+1} \to g^2 + g - 1 = 0. \tag{E.312}$$

The solution of the equation $g^2 + g - 1$ for the growth rate $g_0 = \frac{\sqrt{5}-1}{2} \simeq 0.618$ is the inverse of the famous number $\Phi \equiv \frac{\sqrt{5}+1}{2} = 1/g_0$ (but also $1 + g_0$) known as the golden section or golden ratio which had a great success in the Renaissance era (Pacioli, 1509) under the name of *divina proportione*. It is believed to represent an ideal measure for arts and science (amongst the innumerable examples of golden ratios we have the front face of the Parthenon: the ratio of width and height is given by Φ. Another way to define Φ is to find a part x of a unitary object such that the ratio between 1 (the whole) and x has the same value as the ratio between x and $1 - x$ (what is left). The precise relationship is then

$$\frac{1}{x} = \frac{x}{1 - x}. \tag{E.313}$$

This means that x must obey $x^2 + x - 1 = 0$. Since this equation is the same as eqn E.312 the solution also is the same. $x_0 = \phi = \frac{\sqrt{5}-1}{2} \simeq 0.618$. The inverse of this solution and therefore the left-hand side of eqn E.313 is

$$\frac{1}{x_0} = \frac{\sqrt{5}+1}{2} = \Phi. \tag{E.314}$$

For our purposes, we only need to consider that in the Fibonacci example the growth rate attains a constant value (because of the infinite resources available). This means that

$$\frac{\dot{N}(t)}{N(t)} = g \rightarrow N(t) = N(t_0)e^{g_0(t-t_0)}. \tag{E.315}$$

Whenever we have a constant growth rate g_0 this corresponds to saying that the population increases exponentially with time (when g is positive). Note that if the growth rate were negative (assuming therefore more deaths than births) the total number of individuals would decrease over time.

E.1.2   One species, resources finite

Since in nature 'there is no such thing as a free lunch', it is clear that the above situation is highly simplified. A more realistic assumption is to consider that at a certain point the environment reaches a limit in sustaining the population of the species. This corresponds to fixing a maximum number $\tilde{N}$ of individuals allowed in the ecosystem, this can be obtained by requiring a positive growth rate for $N < \tilde{N}$ and negative for $N > \tilde{N}$.

☛ The simplest assumption that can be made to fulfil this requirement is to put

$$g = g_0[\tilde{N} - N(t)], g_0 > 0. \tag{E.316}$$

With this assumption we have the *logistic equation*

$$\dot{N}(t) = g_0[\tilde{N} - N(t)]N(t). \tag{E.317}$$

This equation has two equilibrium points, one for $N(t) = 0$ and another for $N(t) = \tilde{N}$, The dynamics of evolution reaches the latter one in its steady state. This means that as soon as $N(t_0)$ is larger than 0, any evolution will invariably lead to the limit situation $N(t \to \infty) = \tilde{N}$.

E.1.3   Two species, resources finite

In this case we have the evolution of two species as described by the series of equations for their growth rate

$$
\begin{cases}
\frac{\dot{N_1}(t)}{N_1(t)} &= b_1 - d_1 &= g_1 \\
\frac{\dot{N_2}(t)}{N_2(t)} &= b_2 - d_2 &= g_2
\end{cases}
$$

where one species is dependent upon the other species' success. There are three possibilities.

- For a **prey and predator** interaction we have that an increase in the population of one species will result in a decrease in the number of individuals of the other species and vice versa.

    Therefore without loss of generality if 1 is the predator species we have

$$
\frac{\partial g_1}{\partial N_2} > 0, \quad \frac{\partial g_2}{\partial N_1} < 0. \tag{E.318}
$$

- In the case of **competition** interaction, the two species do not predate each other. Rather they compete for the same finite resources and an increase in population of one species is always related to a decrease in population of the other species.

$$
\frac{\partial g_1}{\partial N_2} < 0, \quad \frac{\partial g_2}{\partial N_1} < 0. \tag{E.319}
$$

- In the case of **mutual** interaction, the two species benefit from each other. Therefore if the population of one species grows, the population of the other species also increases. Using the derivatives also for this case we have the following situation.

$$
\frac{\partial g_1}{\partial N_2} > 0, \quad \frac{\partial g_2}{\partial N_1} > 0. \tag{E.320}
$$

E.1.4   Lotka-Volterra equations

This very important set of equations (Lotka, 1925; Volterra, 1962) refers to the case of competition between species. These equations apply to a more specific situation described by some more hypotheses.

- The birth rate of species 1 (predator) is proportional to the population $N_2(t)$ of prey species. This means $b_1 = \alpha_{12} N_2(t)$ $(\alpha_{12} > 0)$.
- The death rate of species 1 is constant (indicated as $d_1 = \beta_1$).
- The death rate of species 2 (prey) is proportional to the population $N_1(t)$ of predator species. This means $b_2 = \alpha_{21} N_1(t)$ $(\alpha 21 < 0)$.
- The birth rate of species 2 is constant (indicated as $b_2 = \beta_2$).

Under this hypothesis the ecosystem is described by the system of Lotka-Volterra equations (obtained from eqn E.318)

$$\begin{cases} \frac{\dot{N}_1(t)}{N_1(t)} &= \alpha_{12} N_2(t) - \beta_1 \\ \frac{\dot{N}_2(t)}{N_2(t)} &= \alpha_{21} N_1(t) + \beta_2. \end{cases}$$

Only for one precise choice of the parameters is there an (unstable) equilibrium where the two populations remains constant in time. For any other choice of the parameters we have periodic behaviour where the two populations oscillate in time. This nevertheless does not affect the average value of populations. This average value is given by the equilibrium value. If we remove both prey and predators proportionally to their population this results in an increase of prey population.

Bibliography

A copy of the bibliography with links to online documents is available at http://www.citeulike.org/user/gcalda/tag/book.

Adamic, L.A. (1999). The small world web. *Lecture Notes in Computer Sciences*, **1696**, 443–452.

Adamic, L.A. (2002). Zipf, power-laws and Pareto - a ranking tutorial. http://www.hpl.hp.com/research/idl/papers/ranking/ranking.html.

Adamic, L.A. and Huberman, B.A. (2000). Power-law distribution of the World Wide Web. *Science*, **287**, 2115.

Ahnert, S.E., Garlaschelli, D., Fink, T.M., and Caldarelli, G. (2006). An ensemble approach to the analysis of weighted networks. arXiv.org:cond-mat/0604409.

Aiello, W., Chung, F., and Lu, L. (2000). A random graph model for massive graphs. In *Proceedings of the 32nd ACM Symposium on Theory of Computing*, pp. 171–180. ACM Press, New York.

Albert, R., Jeong, H., and Barabási, A.-L. (1999). Diameter of the World Wide Web. *Nature*, **401**, 130–131.

Albert, R., Jeong, H., and Barabási, A.-L. (2000). Error and attack tolerance of complex networks. *Nature*, **406**, 378–382.

Albert, R., Jeong, H., and Barabási, A.-L. (2001). Errata: Error and attack tolerance of complex networks. *Nature*, **409**, 542.

Aldous, D.J. (2001). Stochastic models and descriptive statistics for phylogenetic trees, from Yule to today. *Statistical Science*, **16**, 23–34.

Amaral, L.A.N. and Barthélemy, M. (1999). Small-world networks: Evidence for a crossover picture. *Physical Review Letters*, **82**, 3180–3183.

Anderson, R.M. and May, R.M. (1992). *Infectious diseases of Humans Dynamics and Control*. Oxford University Press, Oxford.

Badger, W.W. (1980). An utility–entropy model for the size distribution of income. In *Mathematical Models as a tool for the social science* (ed. B. West). Gordon and Breach, New York.

Bak, P. and Sneppen, K. (1993). Punctuated equilibrium and criticality in a simple model of evolution. *Physical Review Letters*, **71**, 4083.

Bak, P., Tang, C., and Wiesenfeld, K. (1987). Self-organized criticality. an explanation of $1/f$ noise. *Physical Review Letters*, **59**, 381–384.

Banavar, J.R., Maritan, A., and Rinaldo, A. (1999). Size and form in efficient transportation networks. *Nature*, **399**, 130–132.

Bar-Yossef, Z., Broder, A., Kumar, R., and Tomkins, A. (2004). Sic transit gloria telae: towards an understanding of the web's decay. In *WWW'04: Proceedings of the 13-th international conference on World Wide Web*, pp. 328–337. ACM Press, New York.

Barabási, A.-L. (2002). *Linked: the new science of networks*. Perseus, New York.

Barabási, A.-L. and Albert, R. (1999). Emergence of scaling in random networks. *Science*, **286**, 509–512.

Barabási, A.-L., Jeong, H., Ravasz, E., Néda, Z., Schubert, A., and Vicsek, T. (2002). On the topology of the scientific collaboration networks. *Physica A*, **311**, 590–614.

Baran, P. (1964). *On Distributed Communications*. The Rand Corporation, Memorandum RM-3420-PR.

Barrat, A., Barthélemy, M., Pastor-Satorras, R., and Vespignani, A. (2004a). The architecture of complex weighted networks. *Proceedings of the National Academy of Sciences (USA)*, **101**, 3747–3752.

Barrat, A., Barthélemy, M., Pastor-Satorras, R., and Vespignani, A. (2004b). Weighted evolving networks: Coupling topology and weighted dynamics. *Physical Review Letters*, **92**, 228701.

Barrat, A. and Pastor-Satorras, R. (2005). Rate equation approach for correlations in growing network models. *Physical Review E*, **71**, 036127.

Barrat, A. and Weigt, M. (2000). On the properties of small-world networks. *The European Physical Journal B*, **13**, 547–560.

Barthélemy, M. (2003). Crossover from scale-free to spatial networks. *Europhysics Letters*, **63**, 915–921.

Barthélemy, M. (2004). Betweenness centrality in large complex networks. *The European Physical Journal B*, **38**, 163–168.

Barthélemy, M. and Amaral, L.A.N. (1999a). Small-world networks:evidence for cross over picture. *Physical Review Letters*, **82**, 3180–3183.

Barthélemy, M. and Amaral, L.A.N. (1999b). Erratum: Small-world networks: Evidence for a crossover picture [Phys. Rev. Lett. 82, 3180 (1999)]. *Physical Review Letters*, **82**, 5180–5180.

Batagelj, V. and Mrvar, A. (2000). Some analyses on Erdős collaboration graph. *Social Networks*, **22**, 173–186.

Battiston, S., Bonabeau, E., and Weisbuch, G. (2003). Decision making dynamics in corporate boards. *Physica A*, **322**, 567–582.

Battiston, S. and Catanzaro, M. (2004). Statistical properties of corporate board and director networks. *The European Physical Journal B*, **38**, 345–352.

Becskei, A. and Serrano, L. (2000). Engineering stability in gene networks by autoregulation. *Nature*, **405**, 590–593.

Ben-Jacob, E. and Levine, H. (2001). The artistry of nature. *Nature*, **409**, 985–986.

Benton, M.J. (1993). *The fossil record 2*. Chapman and Hall, London.

Bianconi, G. and Barabási, A.-L. (2001*a*). Bose-einstein condensation in complex networks. *Physical Review Letters*, **86**, 5632–5635.

Bianconi, G. and Barabási, A.-L. (2001*b*). Competition and multiscaling in evolving networks. *Europhysics Letters*, **54**, 436–442.

Bianconi, G., Caldarelli, G., and Capocci, A. (2005). Loops structure of the internet at the autonomous system level. *Physical Review E*, **71**, 066116.

Bianconi, G. and Capocci, A. (2003). Number of loops of size h in growing scale-free networks. *Physical Review Letters*, **90**, 078701.

Boguñá, M. and Pastor-Satorras, R. (2002). Epidemic spreading in correlated complex networks. *Physical Review E*, **66**, 047104.

Boguñá, M. and Pastor-Satorras, R. (2003). Class of correlated random networks with hidden variables. *Physical Review E*, **68**, 036112.

Bollen, J., Rodriguez, M.A., and Van de Sompel, H. (2006). Journal status. *Scientometrics*, **69**. arXiv.org:cs/0601030.

Bollobás, B. (1979). *Graph Theory, An Introductory course*. Springer-Verlag, Berlin.

Bollobás, B. (1985). *Random Graphs*. Academic Press, London.

Bollobás, B. and Riordan, O. (2004). The diameter of a scale-free random graph. *Combinatorica*, **24**, 5–34.

Bonanno, G., Caldarelli, G., Lillo, F., and Mantegna, R.N. (2003). Topology of correlation-based minimal spanning trees in real and model market. *Physical Review E*, **68**, 046130.

Bonanno, G., Caldarelli, G., Lillo, F., Micciché, S., Vandewalle, N., and Mantegna, R.N. (2004). Networks of equities in financial markets. *The European Physical Journal B*, **38**, 363–371.

Börner, K., Dall'Asta, L., Ke, W., and Vespignani, A. (2005). Studying the emerging global brain: Analyzing and visualizing the impact of co-authorship teams. *Complexity*, **10**, 57–67.

Bouchaud, J.-P. and Potters, M. (2000). *Theory of Financial Risk* and Derivatice Pricing: from Statistical Physics to Risk Management. Cambridge University Press, Cambridge.

Brady, R.M. and Ball, R.C. (1984). Fractal growth of copper electrodeposits. Nature, 309, 225–229.

Brandes, U. (2001). A faster algorithm for betweenness centrality. Journal of Mathematical Sociology, 25, 163–177.

Briand, F. and Cohen, J.E. (1984). Community food webs have a scale invariant structure. Nature, 307, 264–266.

Broder, A., Kumar, R., Maghoul, F., Raghavan, P., Rajagopalan, S., Stata, R., Tomkins, A., and Wiener, J.L. (2000). Graph structure in the web. Computer Network, 33, 309–320.

Brown, R. (1828). Microscopical observations of active molecules. Edinburgh New Philosophical Journal, **July-September**, 358–371.

Brun, C., Chevenet, F., Martin, D., Wojcik, J., Guénoche, A., and Jacq, B. (2003). Functional classification of proteins for the prediction of cellular function from a protein-protein interaction network. Genome Biology, 5, R6 1–13.

Bu, T. and Towsley, D. (2002). On distinguishing between internet power-law topology generators. In INFOCOM 2002. Twenty-First Annual Joint Conference of the IEEE Computer and Communications Societies., Volume 2, pp. 638–647. IEEE.

Buchanan, M. (2002). Small World: Uncovering Nature's Hidden Network. Weidenfeld and Nicolson, London.

Burlando, R. (1990). The fractal dimension of taxonomic systems. Journal of Theoretical Biology, 146, 99–114.

Burlando, R. (1993). The fractal geometry of evolution. Journal of Theoretical Biology, 163, 161–172.

Burt, R.S. (1976). Positions in networks. Social Forces, 55, 93–122.

Butland, G., Peregrín-Alvarez, J.M., Li, J., Yang, W., Yang, X., Canadien, V., Starostine, A., Richards, D., Beattie, B., Krogan, N., Davey, M., Parkinson, J., Greenblatt, J., and Emili, A. (2005). Interaction network containing conversed and essential protein complexes in Escherichia coli. Nature, 433, 531–537.

Caldarelli, G. (2001). Cellular model for river networks. Physical Review E, 63, 021118.

Caldarelli, G., Capocci, A., De Los Rios, P., and Muñoz, M.A. (2002). Scale free networks from varying vertex intrinsic fitness. Physical Review Letters, 89, 258702.

Caldarelli, G., Caretta Cartozo, C., De Los Rios, P., and Servedio, V.D.P.

(2004*a*). On the widespread occurence of the inverse square distribution in social sciences and taxonomy. *Physical Review E*, **69**, 035101 (R).

Caldarelli, G., Coccetti, F., and De Los Rios, P. (2004*b*). Preferential exchange: Strengthening connections in complex networks. *Physical Review E*, **70**, 027102.

Caldarelli, G., De Los Rios, P., and Montuori, M. (2001). Statistical features of drainage basins in Mars channel networks. arXiv.org:cond-mat/0107228.

Caldarelli, G., De Los Rios, P., Montuori, M., and Servedio, V.D.P. (2004*c*). Statistical features of drainage basins in Mars channel networks: Can one guess from the landscape the past presence of water? *The European Physical Journal B*, **38**, 387–391.

Caldarelli, G., Giacometti, A., Maritan, A., Rodríguez-Iturbe, I., and Rinaldo, A. (1997). Randomly pinned fluvial landscape evolution. *Physical Review E*, **55**, R4865–R4868.

Caldarelli, G., Higgs, P.G., and McKane, A.J. (1998). Modelling coevolution in multispecies communities. *Journal of Theoretical Biology*, **193**, 345–358.

Caldarelli, G., Marchetti, R., and Pietronero, L. (2000). The fractal properties of Internet. *Europhysics Letters*, **52**, 386–391.

Caldarelli, G. and Servedio, V.D.P. (2006). Topological properties of text networks. In preparation.

Callaway, D.S., Hopcroft, J.E., Kleinberg, J.M., Newman, M.E.J., and Strogatz, S.H. (2001). Are randomly grown graphs really random? *Physical Review E*, **64**, 026118.

Camacho, J., Guimerà, R., and Amaral, L.A.N. (2002). Analytical solution of a model for complex food webs. *Physical Review E*, **65**, 030901.

Campbell, J.Y., Lo, A.W., and MacKinlay, A.C. (1996). *The Econometric of Financial Markets*. Princeton University Press, Princeton.

Caneva, G., Cutini, M., Pacini, A., and Vinci, M. (2002). Analysis of the Colosseum's floristic changes during the last four centuries. *Plant Biosystems*, **136**, 291–311.

Capocci, A., Servedio, V.D.P., Colaiori, F., Buriol, L.S., Donato, D., Leonardi, S., and Caldarelli, G. (2006). Preferential attachment in the growth of social networks: The internet encyclopedia Wikipedia. *Physical Review E*, **74**, 036116.

Capocci, A., Servedio, V. D. P., Caldarelli, G., and Colaiori, F. (2005). Detecting communities in large networks. *Physica A*, **352**, 669–676.

Caretta Cartozo, C., Garlaschelli, D., Ricotta, C., Barthélemy, M., and

Caldarelli, G. (2006). Quantifying the taxonomic diversity in real species communities. arXiv.org:q-bio/0612023.

Carroll, L. (1865). *Alice's adventures in Wonderland*. Macmillan & Co., London.

Caylor, K.K., Scanlon, T.M., and Rodríguez-Iturbe, I. (2004). Feasible optimality of vegetation patterns in river basins. *Geophysical Research Letters*, **31**, L13502.

Cerabino, F. (1994). Mathematicians hit 'Critical Mass' at FAU gathering. *Palm Beach Post*, Florida, March 11.

Champernowne, D. (1953). A model for income distribution. *Economic Journal*, **63**, 318–351.

Chen, P., Xie, H., Maslov, S., and Redner, S. (2007). Finding scientific gems with Google. *Journal of Informetrics*, **1**, XXX. arXiv.org:physics/0604130.

Chien, C., Bartel, P.L., Sternglanz, R., and Fields, S. (1991). The two-hybrid system: A method to identify and clone genes for proteins that interact with a protein of interest. *Proceedings of the National Academy of Sciences (USA)*, **88**, 9578–9582.

Chopard, B., Herrmann, H.J., and Vicsek, T. (1991). Structure and growth mechanisms of mineral dendrites. *Nature*, **353**, 409–412.

Christian, R.R. and Luczkovich, J.J. (1999). Organizing and understanding a winter's seagrass foodweb network through effective trophic levels. *Ecological Modelling*, **117**, 99–124.

Chung, F. and Lu, L. (2001). The diameter of sparse random graphs. *Advances in Applied Mathematics*, **26**, 257–279.

Clauset, A. and Moore, C. (2005). Accuracy and scaling phenomena in internet mapping. *Physical Review Letters*, **94**, 018701.

Cohen, J.E. (1989). *Ecologists' Co-Operative Web Bank. Version 1.00. Machine-readable data base of food webs*. The Rockefeller University New York.

Cohen, J.E., Briand, F., and Newman, C.M. (1990). *Community food webs: data and theory* Biomathematics **20**. Springer-Verlag, Berlin.

Crick, F.H.C. (1958). On protein synthesis. *The Symposia of the Society for Experimental Biology*, **12**, 138–163.

Crick, F.H.C. (1970). Central dogma of molecular biology. *Nature*, **227**, 561–563.

Da Ponte, L. and Mozart, W.A. (1787). Madamina il catalogo è questo. *Don Giovanni K527*, **Act I**, Scene V, Aria nr. 4.

Dall'Asta, L., Alvarez-Hamelin, I., Barrat, A., Vázquez, A., and Vespignani,

A. (2005). Traceroute-like exploration of unknown networks: A statistical analysis. *Lecture Notes in Computer Science*, **3405**, 140–153.

Danon, L., Duch, J., Arenas, A., and Díaz-Guilera, A. (2007). Community structure identification. In *Large Scale Structure and Dynamics of Complex Networks* (ed. G. Caldarelli and A. Vespignani). World-Scientific, Singapore.

Davis, G.F., Yoo, M., and Baker, W. E. (2003). The small world of the American corporate elite, 1982-2001. *Strategic Organization*, **1**, 301–326.

De Castro, R. and Grossman, J.W. (1999). Famous trails to Paul Erdős. *The Mathematical Intelligencer*, **21**, 51–63.

De Masi, G., Iori, G., and Caldarelli, G. (2006). Fitness model for the italian interbank money market. *Physical Review E*, **74**, 066112.

de Solla Price, D.J. (1965). Networks of scientific papers. *Science*, **149**, 510–515.

Deeds, E.J., Ashenberg, O., and Shakhnovich, E.I. (2006). A simple physical model for scaling in protein-protein interaction networks. *Proceedings of the National Academy of Science (USA)*, **103**, 311–316.

Di Battista, G., Patrignani, M., and Pizzonia, M. (2002). Computing the types of the relationships between autonomous systems. Technical Report, Dipartimento di Informatica e Automazione, Università di Roma 3 RT-DIA-73-2002.

Diekmann, O. and Heesterbeek, J.A.P (2000). *Mathematical epidemiology of infectious diseases: model building, analysis and interpretation*. Wiley & Sons, New York.

Diestel, R. (1997-2000-2005). *Graph Theory*. Springer-Verlag, New York, Heidelberg, Berlin.

Dodds, P.S., Muhamad, R., and Watts, D.J. (2003). An experimental study of search in global social networks. *Science*, **301**, 827–829.

Dodds, P.S. and Rothman, D.H. (2000*a*). Geometry of river networks. i. scaling, fluctuations and deviations. *Physical Review E*, **63**, 016115.

Dodds, P.S. and Rothman, D.H. (2000*b*). Geometry of river networks. ii. distribution of component size and number. *Physical Review E*, **63**, 016116.

Dodds, P.S. and Rothman, D.H. (2000*c*). Geometry of river networks. iii. characterization of component connectivity. *Physical Review E*, **63**, 016117.

Donato, D., Millozzi, S., Leonardi, S., and Tsaparas, P. (2005). Mining the inner structure of the web graph. Proceedings of Eighth International Workshop on the Web and Databases (WebDB 2005).

Dorogovtsev, S.N. and Mendes, J.F.F. (2003). *Evolution of Networks From Biological Nets to the Internet and the WWW*. Oxford University Press, Oxford.

Drăgulescu, A. and Yakovenko, V.M. (2001). Exponential and power-law probability distributions of wealth and income in the United Kingdom and the United States. *Physica A*, **299**, 213–221.

Dufrêne, Y.F. (2004). Using nanotechniques to explore microbial surfaces. *Nature Reviews Microbiology*, **2**, 451–460.

Ebel, H., Mielsch, L.-I., and Bornholdt, S. (2002). Scale-free topology of e-mail networks. *Physical Review E*, **66**, 035103.

Eckmann, J. P., Moses, E., and Sergi, D. (2004). Entropy of dialogues creates coherent structures in e-mail traffic. *Proceeding of the National Academy of Science (USA)*, **101**, 14333–14337.

Eldredge, N. and Gould, S.J. (1973). Punctuated equilibria: an alternative to phyletic gradualism. In *Models in Paleobiology* (ed. T. Schopf), pp. 82–115. Freeman & Cooper, San Francisco.

Elton, C.S. (1927). *Animal Ecology*. Sidgwick & Jackson, London. Reprinted by Chicago University Press (2001).

Engel, A. and Müller, D.J. (2000). Observing single biomolecules at work with the atomic force microscope. *Nature Structural Biology*, **7**, 715–718.

Erdős, P. and Rényi, A. (1959). On random graphs. *Publicationes Mathematicae Debrecen*, **6**, 290–297.

Erdős, P. and Rényi, A. (1960). On the evolution of random graphs. *Publications of the Mathematical Institute of the Hungarian Academy of Sciences*, **5**, 17–61.

Erdős, P. and Rényi, A. (1961). On the strength of connectedness of a random graph. *Acta Mathematica Hungarica*, **12**, 261–267.

Euler, L. (1736). Solutio problematis ad geoetriam situs pertinentis. *Commentarii Academiae Scientiarum Imperialis Petropolitanae*, **8**, 128–140.

F. Schepisi (Director) and J. Guare (Writer) (1993). *Six degrees of separation*. Metro Goldwyn Mayer.

Faloutsos, M., Faloutsos, P., and Faloutsos, C. (1999). On power-law relationships of the internet topology. *Proc. ACM SIGCOMM, Computer Communication Review*, **29**, 251–262.

Farahat, A., LoFaro, T., Miller, J.C., Rae, G., and Ward, L.A. (2006). Authority rankings from HITS, PageRank, and SALSA: Existence, uniqueness, and effect of initialization. *SIAM Journal of Scientific Computing*, **27**, 1181–1201.

Feller, W. (1968). *An Introduction to Probability Theory and its Applications*. Wiley & Sons, New York.

Ferrer i Cancho, R. and Solé, R.V. (2001). The small world of human language. *Proceedings of the Royal Society B, London*, **268**, 2261–2265.

Ferrer i Cancho, R. and Solé, R.V. (2003). Optimisation in complex networks. *Lectures Notes in Physics*, **625**, 114–126.

Fink, T. (2006). *Man's book*. Weidenfeld and Nicholson, London.

Flory, P.J. (1953). *Principles of Polymer Chemistry*. Cornell University Press, Ithaca NY (USA).

Formstecher, E., Aresta, S., Collura, V., Hamburger, A., Meil, A., Trehin, A., Reverdy, C., Betin, V., Maire, S., Brun, C., Jacq, B., Arpin, M., Bellaiche, Y., Bellusci, S., Benaroch, P., Bornens, M., Chanet, R., Chavrier, P., Delattre, O., Doye, V., Fehon, R., Faye, G., Galli, T., Girault, J. A., Goud, B., de Gunzburg, J., Johannes, L., Junier, M. P., Mirouse, V., Mukherjee, A., Papadopoulo, D., Perez, F., Plessis, A., Rosse, C., Saule, S., Stoppa-Lyonnet, D., Vincent, A., M, M. White, Legrain, P., Wojcik, J., Camonis, J., and Daviet, L. (2005). Protein interaction mapping: A drosophila case study. *Genome Research*, **15**, 376–374.

Förster, J., Famili, I., Fu, P., Palsson, B.Ø., and Nielsen, J. (2003). Genome-scale reconstruction of the saccharomyces cerevisiae metabolic network. *Genome Research*, **13**, 244–253.

Freeman, L.C. (1977). A set of measures of centrality based upon betweenness. *Sociometry*, **40**, 35–41.

Fromont-Racine, M., Rain, J.-C., and Legrain, P. (1997). Toward a functional analysis of the yeast genome through exhaustive two-hybrid screens. *Nature Genetics*, **16**, 277 – 282.

Fronczak, A., Fronczak, P., and Holyst, J.A. (2003). Mean-field theory for clustering coefficient in Barabási-Albert networks. *Physical Review E*, **68**, 046126.

Gabaix, X. (1999). Zipf's law for cities: an explanation. *Quarterly Journal of Economics*, **114**, 739–767.

Gabrielli, A. (2000). Private communication.

Gao, L. (2001). On inferring autonomous system relationships in the internet. *IEEE/ACM Transaction on Networking*, **9**, 733–745.

Garlaschelli, D., Battiston, S., Castri, M., Servedio, V.D.P., and Caldarelli, G. (2005). The scale free topology of market investments. *Physica A*, **350**, 491–499.

Garlaschelli, D., Caldarelli, G., and Pietronero, L. (2003). Universal scaling relations in food webs. *Nature*, **423**, 165–168.

Garlaschelli, D. and Loffredo, M.I. (2004*a*). Fitness-dependent topological properties of the world trade web. *Physical Review Letters*, **93**, 188701.

Garlaschelli, D. and Loffredo, M.I. (2004*b*). Patterns of link reciprocity in directed networks. *Physical Review Letters*, **93**, 268701.

Garlaschelli, D. and Loffredo, M.I. (2005). Structure and evolution of the world trade network. *Physica A*, **355**, 138–144.

Gibrat, R. (1930). Une loi des répartitions économiques: l'effet proportionnel. *Bulletin de la Statistique Générale de la France*, **19**, 469–514. described in M. Armatte "Robert Gibrat et la loi de l'effet proportionnel" *Mathématiques et Sciences Humaines*, **129** 5-35 (1995).

Gibrat, R. (1931). *Les inégalités économiques*. Libraire du Recueil Siray, Paris.

Giot, L., Bader, J. S., Brouwer, C., Chaudhuri, A., Kuang, B., Li, Y., Hao, Y. L., Ooi, C. E., Godwin, B., Vitols, E., Vijayadamodar, G., Pochart, P., Machineni, H., Welsh, M., Kong, Y., Zerhusen, B., Malcolm, R., Varrone, Z., Collis, A., Minto, M., Burgess, S., McDaniel, L., Stimpson, E., Spriggs, F., Williams, J., Neurath, K., Ioime, N., Agee, M., Voss, E., Furtak, K., Renzulli, R., Aanensen, N., Carrolla, S., Bickelhaupt, E., Lazovatsky, Y., DaSilva, A., Zhong, J., Stanyon, C. A., Finley Jr., R. L., White, K. P., Braverman, M., Jarvie, T., Gold, S., Leach, M., Knight, J., Shimkets, R. A., McKenna, M. P., Chant, J., and Rothberg, J. M. (2003). A Protein Interaction Map of Drosophila melanogaster. *Science*, **302**, 1727–1736.

Girvan, M. and Newman, M.E.J. (2002). Community structure in social and biological networks. *Proceeding of the National Academy of Science (USA)*, **99**, 7821–7826.

Gnedenko, B. V. (1962). *The theory of probability*. Chelsea, New York.

Goh, K.-I., Kahng, B., and Kim, D. (2001). Universal behaviour of load distribution in scale-free networks. *Physical Review Letters*, **87**, 278701.

Goldstein, M.L., Morris, S.A., and Yen, G.G. (2004). Problems with fitting to power-law distribution. *The European Physical Journal B*, **41**, 255–258.

Goldwasser, L. and Roughgarden, J. (1997). Sampling effects and the estimation of food-web properties. *Ecology*, **78**, 41–54.

Golub, G. and Van Loan, C.F. (1989). *Matrix Computation*. Johns Hopkins University Press, Baltimore MD.

González, M.C., Lind, P.G., and Herrmann, H.J. (2006). System of mobile agents to model social networks. *Physical Review Letters*, **96**, 088702.

Gould, S.J. and Eldredge, N. (1977). Punctuated equilibria. *Paleobiology*, **3**, 115–151.

Gower, J.C. (1966). Some distance properties of latent root and vector methods used in multivariate analysis. *Biometrika*, **53**, 325–338.

Goyal, S., van der Leij, M., and Moraga-Gonzàles, J.L. (2004). Economics: an emerging small world? Working Papers 2004.84, Fondazione Eni Enrico Mattei. available at http://ideas.repec.org/p/fem/femwpa/2004.84.html.

Green, J.E. and Moore, M.A. (1982). Application of directed lattice animal theory to river networks. *Journal of Physics A*, **15**, L597–L599.

Greuter, W., McNeill, J., Barrie, F. R., Burdet, H.-M., Demoulin, V., Filgueiras, T. S., Nicolson, D. H., Silva, P. C., Skog, J. E., Trehane, P., Turland, N. J., and Hawksworth, D. L. (2000). *International Code of Botanical Nomenclature (St Louis Code). Regnum Vegetabile 138.* Koeltz Scientific Books, Königstein.

Grossman, J.W. (2003). The Erdős number project. http://www.oakland.edu/enp.

Grossman, J.W. and Ion, P.D.F. (1995). On a portion of the well known collaboration graph. *Congressus Numerantium*, **108**, 129–131.

Guimerà, R., Danon, L., Díaz-Guilera, A., Giralt, F., and Arenas, A. (2002). Self-similar community structure in organisations. *Physical Review E*, **68**, 065103.

Guimerà, R., Sales-Pardo, M., and Amaral, L.N.A. (2004). Modularity from fluctuations in random graphs and complex networks. *Physical Review E*, **70**, 025101.

Gutenberg, B. and Richter, C.F. (1955). Magnitude and energy of earthquakes. *Nature*, **176**, 795–795.

Hack, J.T. (1957). Studies of longitudinal stream profiles in Virginia and Maryland. *United States Geological Survey Professional Paper*, **294-B**, 45–97.

Hack, J.T. (1965). Geomorphology of the Shenandoah Valley, Virginia and West Virginia, and origin of the residual ore deposits. *United States Geological Survey Professional Paper*, **484**.

Hall, K.M. (1970). An r-dimensional quadratic placement algorithm. *Management Science*, **17**, 219–229.

Hall, S.J. and Raffaelli, D.J. (1991). Static patterns in food webs: lessons from a large web. *Journal of Animal Ecology*, **60**, 823–842.

Han, J.-D. J., Bertin, N., Hao, T., Goldberg, D. S., Berriz, G. F., Zhang, L. V., Dupuy, D., Walhout, A. J. M., Cusick, M. E., Roth, F. P., and Vidal, M. (2004). Evidence for dynamically organized modularity in the yeast protein-protein interaction network. *Nature*, **430**, 88–93.

Havens, K. (1992). Scale and structure in natural food webs. *Science*, **257**,

1107–1109.

Head III, J.W., Hiesinger, H., Ivanov, M.A., Kreslavsky, M.A., Pratt, S., and Thomson, B.J. (1999). Possible ancient oceans on Mars: evidence from Mars orbiter laser altimeter data. *Science*, **286**, 2134.

Hoffmann, P. (1998). *The man who loved only numbers*. Hyperion, Boston.

Holloway, T., Božičević, M., and Börner, K. (2005). Analyzing and visualizing the semantic coverage of wikipedia and its authors. arXiv:cs.IR/0512085.

Holme, Petter (2006). Detecting degree symmetries in networks. *Physical Review E (Statistical, Nonlinear, and Soft Matter Physics)*, **74**, 036107.

Howard, A.D. (1994). A detachment-limited model of drainage basin evolution. *Water Resources Research*, **30**, 2261–2286.

Huberman, B.A. and Adamic, L.A. (1999). Internet: Growth dynamics of the world-wide web. *Nature*, **401**, 131–131.

Huffaker, B., Fomenkov, M., Plummer, D., Moore, D., and Claffy, K. (2002). Distance metrics in the internet. Presented at IEEE International Telecommunications Symposium,.

Huggins, M.L. (1942). Thermodynamic properties of solutions of long-chain compounds. *Annals N. Y. Academy of Science*, **43**, 1–32.

Hurst, H.E. (1951). Long term storage capacity in reservoirs. *Transactions of the American Society of Civil Engeneering*, **116**, 770–799.

Huxham, M., Beaney, S., and Raffaelli, D. (1996). Do parasites reduce the chances of triangulation in a real food web? *Oikos*, **76**, 284–300.

Ito, T., Chiba, T., Ozawa, R., Yoshida, M., Hattori, M., and Sakaki, Y. (2001). A comprehensive two-hybrid analysis to explore the yeast protein interactome. *Proceeding of the National Academy of Science (USA)*, **98**, 4569–4574.

Jackson, J.D. (1998). *Classical Electrodynamics, 3rd ed.* Wiley & Sons, New York.

Jeong, H., Mason, S.P., Barabási, A.-L., and Oltvai, Z.N. (2001). Lethality and centrality in protein networks. *Nature*, **411**, 41–42.

Jeong, H., Tombor, B., Albert, R., Oltvai, Z.N., and Barabási, A.-L. (2000). The large-scale organization of metabolic networks. *Nature*, **407**, 651–654.

Kaplan, A. (1955). An experimental study of ambiguity and context. *Mechanical Translation*, **2**, 39–46.

Kapteyn, J.C. (1903). *Skew Frequency curves in Biology and Statistics*. Astronomical Laboratory, Noordhoff Groeningen, The Netherlands.

Kapteyn, J. C. (1918). Skew frequency curves in biology and statistics. *Molecular and General Genetics MGG*, **19**, 205–206.

Kauffman, S.A. (1969). Metabolic stability and epigenesis in randomly constructed genetic nets. *Journal of Theoretical Biology*, **22**, 434–467.

Kauffman, S.A. (1993). *The Origin of Order*. Oxford University Press, Oxford.

Kholodenko, B. N., Kiyatkin, A., Bruggeman, F. J., Sontag, E., Westerhoff, H. V., and Hoek, J. B. (2002). Untangling the wires: A strategy to trace functional interactions in signaling and gene networks. *Proceedings National Academy of Science (USA)*, **99**, 12841–12846.

Kleinberg, J.M. (1999a). Authoritative sources in a hyperlinked environment. *Journal of the ACM*, **46**, 604–632.

Kleinberg, J.M. (1999b). Hubs, authorities and communities. *ACM Computing Surveys*, **31**(4es, Article no.5), 1–5.

Kleinberg, J.M. and Lawrence, S. (2001). Network analysis: The structure of the web. *Science*, **294**, 1849–1850.

Klemm, K. and Eguíluz, V.M. (2002). Highly clustered scale-free networks. *Physical Review E*, **65**, 036123.

Krapivski, P.L., Rodgers, G.J., and Redner, S. (2001). Degree distributions of growing networks. *Physical Review Letters*, **86**, 5401–5404.

Kumar, R., Raghavan, P., Rajagopalan, S., Sivakumar, D., Tompkins, A., and Upfal, E. (2000). The web as a graph. In *PODS '00: Proceedings of the nineteenth ACM SIGMOD-SIGACT-SIGART symposium on Principles of database systems*, New York, pp. 1–10. ACM Press.

Kumar, R., Raghavan, P., Rajagopalan, S., and Tomkins, A. (1999a). Extracting large-scale knowledge bases from the web. In *VLDB '99: Proceedings of the 25th International Conference on Very Large Data Bases*, San Francisco, pp. 639–650. Morgan Kaufmann Publishers Inc.

Kumar, R., Raghavan, P., Rajagopalan, S., and Tomkins, A. (1999b). Trawling the web for emerging cyber-communities. *Computer Networks*, **31**, 1481–1493.

Lahèrrere, J. and Sornette, D. (1998). Stretched exponential distributions in nature and economy: fat tails with characteristic scales. *The European Physical Journal B*, **2**, 525–539.

Lakhina, A., Byers, J.W., Crovella, M., and Matta, I. (2003). On the geographic location of internet resources. *IEEE Journal on Selected Areas in Communications*, **21**, 934–948.

Laloux, L., Cizeau, P., Bouchaud, J.-P., and Potters, M. (1999). Noise dressing of financial correlation matrices. *Physical Review Letters*, **83**, 1467–1470.

Langville, A.N. and Meyer, C.D. (2005). A survey of eigenvector methods of web information retrieval. *The SIAM Review*, **47**, 135–161.

Langville, A.N. and Meyer, C.D. (2006). *Google's PageRank and Beyond*. Princeton University Press, Princeton.

Laub, M.T., McAdams, H.H., Feldblyum, T., Fraser, C.M., and Shapiro, L. (2000). Global analysis of the genetic network controlling a bacterial cell cycle. *Science*, **290**, 2144–2148.

Laumann, E.O., Gagnon, J.H., Michael, R.T., and Michaels, S. (1994). *The Social Organization of Sexuality*. Univeristy of Chicago Press, Chicago.

Laura, L., Leonardi, S., Millozzi, S., Meyer, U., and Sibeyn, J.F. (2003). Algorithms and experiments for the web graph. *Lecture Notes in Computer Science*, **2832**, 703–714.

Lawrence, S. and Giles, C.L. (1998). Searching the world wide web. *Science*, **280**, 98–100.

Lawrence, S. and Giles, C.L. (1999). Accessibility of information on the web. *Nature*, **400**, 107–109.

Lehmann, S., Lautrup, B., and Jackson, A.D. (2003). Citation networks in high energy physics. *Physical Review E*, **68**, 026113.

Leicht, E.A., Holme, P., and Newman, M.E.J. (2006). Vertex similarity in networks. *Physical Review E*, **73**, 026120.

Lewin, B. (1998). *Sex i Sverige. Om Sexuallivet i Sverige 1996 (Sex in Sweden. On the Sexual life in Sweden 1996)*. National Institute of Public Health, Stockholm.

Li, X., Jin, Y.Y., and Chen, G. (2003). Complexity and synchronization of the world trade web. *Physica A*, **328**, 287–296.

Liljeros, F., Edling, C.R., Amaral, L.A.N., Stanley, H.E., and Åberg, Y. (2001). The web of human sexual contacts. *Nature*, **411**, 907–908.

Lotka, A.J. (1925). *Elements of Physical Biology*. Williams and Wilkins, Baltimore.

Ma, H.-W. and Zeng, A.-P. (2003). Reconstruction of metabolic networks from genome data and analysis of their global structure for various organisms. *Bioinformatics*, **19**, 1423–1430.

Mahadevan, L. and Rica, S. (2005). Self-organized origami. *Science*, **307**, 1740.

Mahadevan, P., Krioukov, D., Fomenkov, M., Huffaker, B., Dimitropoulos, X., kc claffy, and Vahdat, A. (2005). Lessons from three views of the Internet topology. arXiv.org:cs.NI/0508033.

Malin, M.C. and Edgett, K.S. (2000a). Evidence for recent groundwater seepage and surface runoff on Mars. *Science*, **288**, 2330–2335.

Malin, M.C. and Edgett, K.S. (2000b). Sedimentary rocks of early Mars.

Science, **290**, 1927–1937.

Mandelbrot, B. (1953). An informational theory of the statistical structure of languages. In *Communication Theory* (ed. W. Jackson), pp. 486–504. Betterworth, London.

Mandelbrot, B. (1963). The variation of certain speculative prices. *Journal of Business*, **36**, 394–419.

Mandelbrot, B. (1975). *Les Objets Fractals, Form, Hasard et Dimension.* Flammarion, Paris.

Mandelbrot, B. (1983). *The Fractal Geometry of Nature.* Freeman & Cooper, San Francisco.

Mangan, S. and Alon, U. (2003). Structure and function of the feed-forward loop network motif. *Proceeding of the National Academy of Science (USA)*, **100**, 11980–11985.

Manrubia, S.C. and Zanette, D.H. (1998). Intermittency model for urban development. *Physical Review E*, **58**, 295–302.

Mantegna, R.N. (1999). Hierarchical structure in financial markets. *The European Physical Journal B*, **11**, 193–197.

Marani, M., Banavar, J.R., Caldarelli, G., Maritan, A., and Rinaldo, A. (1998). Stationary self-organized fractal structures in an open, dissipative electrical system. *Journal of Physics A*, **31**, L337–L343.

Mardia, K.V., Kent, J.T., and Bibby, J.M. (1980). *Multivariate Analysis.* Academic Press, San Diego. reprinted by Elsevier (2006).

Maritan, A., Rinaldo, A., Rigon, R., Giacometti, A., and Rodríguez-Iturbe, I. (1996). Scaling laws for river networks. *Physical Review E*, **53**, 1510–1515.

Markowitz, H. (1952). Portfolio selection. *The Journal of Finance*, **7**, 77–91.

Martinez, N.D. (1991). Artifacts or attributes? effects of resolution on the Little Rock Lake food web. *Ecological Monographs*, **61**, 367–392.

Martinez, N.D. (1992). Constant connectance in community food webs. *American Naturalist*, **139**, 1208–1218.

Martinez, N.D., Hawkins, B.A., Dawah, H.A., and Feifarek, B.P. (1999). Effects of sampling effort on characterization of food-web structure. *Ecology*, **80**, 1044–1055.

Martinez, N.D. and Lawton, J.H. (1995). Scale and food-web structure–from local to global. *OIKOS*, **73**, 148–154.

Maslov, S. and Sneppen, K. (2002). Specifity and stability in topology of protein networks. *Science*, **296**, 910–913.

Matsushita, M. and Fujikawa, H. (1990). Diffusion limited growth in bacterial colony formation. *Physica A*, **168**, 498–506.

Mazurie, A., Bottani, S., and Vergassola, M. (2005). An evolutionary and functional assessment of regulatory network motifs. *Genome Biology*, **6**, R35.

McAlister, D. (1879). The law of the geometrical mean. *Proceedings of the Royal Society, London*, **29**, 367–376.

Memmott, J., Martinez, N.D., and Cohen, J.E. (2000). Predators, parasites and pathogens: species richness, trophic generality, and body sizes in a natural food web. *Animal Ecology*, **69**, 1–15.

Merton, R.K. (1968). The Matthew effect in science: The reward and communication systems of science are considered. *Science*, **159**, 56–63.

Merton, R.K. (1988). The Matthew effect in science, II: Cumulative advantage and the symbolism of intellectual property. *ISIS*, **79**, 606–623.

Metropolis, N., Rosembluth, A.W., Rosembluth, M.N., Teller, A.H., and Teller, E. (1953). Equation of state calculations by fast computing machines. *Journal of Chemical Physics*, **21**, 1087–1092.

Milgram, S. (1967). The small world problem. *Psychology Today*, **1**, 60–67.

Miller, J.C., Rae, G., Schaefer, F., Ward, L.A., Lofaro, T., and Farahat, A. (2001). Modifications of kleinberg's hits algorithm using matrix exponentiation and web log records. In *SIGIR '01: Proceedings of the 24th annual international ACM SIGIR conference on Research and development in information retrieval*, pp. 444–445. ACM Press, New York.

Milo, R., Shen-Orr, S., Itzkovitz, S., Kashtan, N., Chklovskii, D., and Alon, U. (2002). Network motifs: Simple building blocks of complex networks. *Science*, **298**, 824–827.

Mitzenmacher, M. (2004). A brief history of generative models for power-law and log-normal. *Internet Mathematics*, **1**, 226–251.

Montoya, J.M. and Solé, R.V. (2002). Small world patterns in food webs. *Journal of Theoretical Biology*, **214**, 405–412.

Morita, S. (2006). Crossovers in scale free networks on geographical space. *Physical Review E*, **73**, 035104(R).

Nagle, F.F. (1974). Statistical mechanics of the melting transition in lattice models of polymers. *Proceedings of the Royal Society A, London*, **337**, 569–589.

Nelson, D.L. and Cox, M.M. (2005). *Lehninger Principles of Biochemistry*. Freeman & Cooper New York.

Nelson, D.L., McEvoy, C.L., and Schreiber, T.A. (1998). *The University of South Florida word association, rhyme, and word fragment norms*. USF, http://www.usf.edu/FreeAssociation/.

New American Bible (2002). United States Conference of Catholic Bishops

3211 4th Street, N.E., Washington DC.

Newman, M.E.J. (2001*a*).   Scientific collaboration networks: I. network construction and fundamental results. *Physical Review E*, **64**, 016131.

Newman, M.E.J. (2001*b*).   Scientific collaboration networks: II. Shortest paths, weighted networks, and centrality. *Physical Review E*, **64**, 016132.

Newman, M.E.J. (2001*c*). The structure of scientific collaboration networks. *Proceedings of the National Academy of Science (USA)*, **98**, 404–409.

Newman, M.E.J. (2002*a*). Assortative mixing in networks. *Physical Review Letters*, **89**, 208701.

Newman, M.E.J. (2002*b*). The structure and function of networks. *Computer Physics Communications*, **147**, 40–45.

Newman, M.E.J. (2003). The structure and function of complex networks. *SIAM Review*, **45**, 167–256.

Newman, M.E.J. (2004*a*).   Coauthorship networks and patterns of scientific collaboration. *Proceedings of the National Academy of Science (USA)*, **101**, 5200–5205.

Newman, M.E.J. (2004*b*). Detecting community structure in networks. *The European Physical Journal B*, **38**, 321–330.

Newman, M.E.J. (2005).   Power laws Pareto distributions and Zipf's laws. *Contemporary Physics*, **46**, 323–351.

Newman, M.E.J. (2006). Modularity and community structure in networks. *Proceedings of the National Academy of Science (USA)*, **103**, 8577–8582.

Newman, M.E.J. and Girvan, M. (2003). Finding and evaluating community structure in networks. *Physical Review E*, **69**, 026113.

Newman, M.E.J., Moore, C., and Watts, D.J. (2000). Mean field solution of the small-world network model. *Physical Review Letters*, **84**, 3201–3204.

Newman, M.E.J. and Sibani, P. (1999). Extinction, diversity and survivorship of taxa in the fossil record. *Proceedings of the Royal Society B, London*, **266**, 1593–1599.

Niemeyer, L., Pietronero, L., and Wiesmann, H.J. (1984). Fractal dimension of dielectric breakdown. *Physical Review Letters*, **52**, 1033–1036.

Onnela, J.-P, Chakraborti, A., Kaski, K., Kertész, J., and Kanto, A. (2003).   Dynamics of market correlations: Taxonomy and portfolio analysis. *Physical Review E*, **68**, 056110.

Onnela, J.-P, Kaski, K., and Kertész, J. (2004). Clustering and information in correlation based financial networks. *The European Physical Journal B*, **38**, 353–362.

Onnela, J.-P, Samaräki, J., Kertész, J., and Kaski, K. (2005).   Intensity and coherence of motifs in weighted complex networks. *Physical Review*

E, **71**, 065103(R).

Pacioli, L. (1509). De divina proportione. Silvana Editoriale, Milan.

Page, L., Brin, S., Motwami, R., and Winograd, T. (1999). The pagerank citation ranking: bringing order to the web. Technical Report, Computer Science Department, Stanford University.

Palla, G., Derényi, I., Farkas, I., and Vicsek, T. (2005). Uncovering the overlapping community structure of complex networks in nature and society. *Nature*, **435**, 814–818.

Papin, J.A., Price, N.D., Wiback, S.J., Fell, D.A., and Palsson, B.Ø. (2003). Metabolic pathways in the post-genome era. *Trends in Biochemical Science*, **28**, 250–258.

Pareto, V. (1897). *Course d'économie politique*. F. Pichou, Lausanne and Paris.

Pastor-Satorras, R., Vázquez, A., and Vespignani, A. (2001). Dynamical and correlation properties of the internet. *Physical Review Letters*, **87**, 258701.

Pastor-Satorras, R. and Vespignani, A. (2002). Immunization of complex networks. *Physical Review E*, **65**, 036104.

Pastor-Satorras, R. and Vespignani, A. (2004). *Evolution and Structure of Internet: A Statistical Physics Approach*. Cambridge University Press, Cambridge.

Pauly, D., Christensen, V., Dalsgaard, J., Froese, R., and Torres Jr, F (1998). Fishing down marine food webs. *Science*, **279**, 860–863.

Petermann, T. and De Los Rios, P. (2004). Exploration of scale-free networks: Do we measure the real exponents? *The European Physical Journal B*, **38**, 201–204.

Pimm, S.L. (1991). *The Balance of Nature? Ecological Issues in the Conservation of Species and Communities*. University of Chicago Press, Chicago.

Pimm, S.L., Lawton, J.H., and Cohen, J.E. (1991). Food webs patterns and their consequences. *Nature*, **350**, 669–674.

Plerou, V., Gopikrishnan, P., Rosenow, B., Amaral, L.A.N., and Stanley, H.E. (1999). Universal and non-universal properties of cross correlations in financial time series. *Physical Review Letters*, **83**, 1471–1474.

Polis, G.A. (1991). Complex trophic interactions in deserts: an empirical critique of food-web theory. *American Naturalist*, **138**, 123–155.

Pothen, A., Simon, H.D., and Liou, K.-P. (1990). Partitioning sparse matrices with eigenvectors of graphs. *SIAM Journal of Matrix Analysis and Applications*, **11**, 430–452.

Radicchi, F., Castellano, C., Cecconi, F., Loreto, V., and Parisi, D. (2004). Defining and identifying communities in networks. *Proceedings of the National Acadamy of Science (USA)*, **101**, 2658–2663.

Ravasz, E. and Barabási, A.-L. (2003). Hierarchical organization in complex networks. *Physical Review E*, **67**, 026112.

Ravasz, E., Somera, A.L., Mongru, D.A., Oltvai, Z.N., and Barabási, A.-L. (2002). Hierarchical organization of modularity in metabolic networks. *Science*, **297**, 1551–1555.

Redner, S. (1998). How popular is your paper? An empirical study of the citation distribution. *The European Physical Journal B*, **4**, 131–134.

Reed, W.J. and Hughes, B.D. (2002). From genes families and genera to incomes and internet file sizes: why power-laws are so common in nature. *Physical Review E*, **66**, 067103.

Reichl, L.E. (1980). *A Modern Course in Statistical Physics* (2nd edn). E. Arnold, London. Wiley-Interscience (1998).

Rinaldo, A., Maritan, A., Colaiori, F., Flammini, A., Rigon, R., Rodríguez-Iturbe, I., and Banavar, J.R. (1996). Thermodynamics of fractal networks. *Physical Review Letters*, **76**, 3364–3367.

Rinaldo, A., Rodríguez-Iturbe, I., Rigon, R., Ijjasz-Vasquez, E., and Bras, R.L. (1993). Self-organized fractal river networks. *Physical Review Letters*, **70**, 822–825.

Ristroph, L., Thrasher, M., Mineev-Weinstein, M., and Swinney, H.L. (2006). Fjords in viscous fingering: selection of width and opening angle. *Physical Review E*, **74**, 015201(R).

Rodríguez-Iturbe, I., Ijjasz-Vasquez, E., Bras, R.L., and Tarboton, D.G. (1992*a*). Power-law distributions of discharge mass and energy in river basins. *Water Resources Research*, **28**, 1089–1093.

Rodríguez-Iturbe, I. and Rinaldo, A. (1996). *Fractal River Basins: Chance and Self-Organization*. Cambridge University Press, Cambridge.

Rodríguez-Iturbe, I., Rinaldo, A., Rigon, R., Bras, R.L., Ijjasz-Vasquez, E., and Marani, A. (1992*b*). Fractal structures as least energy patterns: the case of river networks. *Geophysical Research Letters*, **19**, 889–892.

Rogers, I. (2005). The google pagerank algorithm and how it works. http://iprcom.com/papers/pagerank/.

Roget, P.M. (2002). *Roget's Thesaurus of English words and phrases:150th Anniversary Edition (Reprint Edited by G.W. Davidson of the 1852 Edition)*. Penguin Books Ltd, London.

Rossini, G.P. and Camellini, L. (1994). Oligomeric structure of cytosoluble estrogen-receptor complexes as studied by anti-estrogen receptor antibodies and chemical crosslinking of intact cells. *Journal of Steroid Biochemistry, Molecular Biology*, **50**, 241–252.

Scheidegger, A.E. (1970). Stochastic models in hydrology. *Water Resources Research*, **6**, 750–755.

Seary, A.J. and Richards, W.D. (1995). Partitioning networks by eigenvectors. *Proceedings of the International Conference on Social Networks*, **1: Methodology**, 47–58.

Segrè, D., Vitkup, D., and Church, G.M. (2002). Analysis of optimality in natural and perturbed metabolic networks. *Proceedings of the National Academy of Science (USA)*, **99**, 15112–15117.

Sepkoski Jr, J.J. (1992). A compendium of fossil marine animal families. *Milwaukee Public Museum Contribution in Biology and Geology*, **83**, 1–156. (2-nd edition).

Sepkoski Jr, J.J. (1993). Ten years in the library: new data confirm paleontological patterns. *Paleobiology*, **19**, 43–51.

Serrano, M.Á. and Boguñá, M. (2003). Topology of the world trade web. *Physical Review E*, **68**, 015101 (R).

Servedio, V.D.P., Caldarelli, G., and Buttà, P. (2004). Vertex intrinsic fitness: how to produce arbitrary scale-free networks. *Physical Review E*, **70**, 056126.

Simon, H.A. (1955). On a class of skew distribution function. *Biometrika*, **42**, 425–440.

Simon, H.A. (1960). Some further notes on a class of skew distribution function. *Information and Control*, **3**, 80–88.

Stanley, H.E. (1971). *Introduction to Phase Transitions and Critical Phenomena*. Oxford University Press, Oxford.

Stauffer, D. and Aharony, A. (1992). *Introduction to Percolation Theory* (2nd edn). Taylor and Francis.

Stepinski, T.F., Marinova, M.M., McGovern, P.J., and Clifford, S.M. (2002). Fractal analysis of drainage basins on mars. *Geophysical Research Letters*, **29**, 1189.

Steyvers, M. and Tenenbaum, J.B. (2005). The large scale structure of semantic networks: Statistical analyses and a model of semantic growth. *Cognitive Science*, **29**, 41–78.

Stouffer, D.B., Camacho, J., Guimerà, R., Ng, C.A., and Amaral, L.A.N. (2005). Quantitative patterns in the structure of model and empirical food webs. *Ecology*, **86**, 1301–1311.

Sugihara, G., Schoenly, K., and Trombla, A. (1989). Scale invariance in food web properties. *Science*, **245**, 48–52.

Szabó, G., Alava, M.J., and Kertész, J. (2002). Shortest paths and load scaling in scale-free trees. *Physical Review E*, **66**, 026101.

Szabó, G., Alava, M.J., and Kertész, J. (2004). Clustering in complex networks. *Lecture Notes in Physics*, **650**, 139–162. Complex Networks (E. Ben-Naim, H. Frauenfelder, Z. Toroczkai Eds.).

Tarboton, D.G., Bras, R.L., and Rodríguez-Iturbe, I. (1988). The fractal nature of river networks. *Water Resources Research*, **24**, 1317–1322.

Travers, J. and Milgram, S. (1969). An experimental study of the small world problem. *Sociometry*, **32**, 425–443.

Tsallis, C. and de Albuquerque, M.P. (1999). Are citations of scientific papers a case of non extensivity? *The European Physical Journal B*, **13**, 777–780.

Tumminello, M., Aste, T., Di Matteo, T., and Mantegna, R.N. (2005). A tool for filtering information in complex systems. *Proceeding of the National Academy of Science (USA)*, **102**, 10421–10426.

Tyler, J.R., Wilkinson, D.M., and Huberman, B.A. (2003). E-mail as a spectroscopy: Automated discovery of community structure within organizations. In *Communities and Technologies*, pp. 81–96. Kluwer, Deventer.

Uetz, P., Giot, L., Cagney, G., Mansfield, T. A., Judson, R. S., Knight, J.R., Lockshon, D., Narayan, V., Srinivasan, M., Pochart, P., Qureshi-Emili, A., Li, Y., Godwin, B., Conover, D., Kalbfleisch, T., Vijayadamodar, G., Yang, M., Johnston, M., Fields, S., and Rothberg, J. M. (2000). A comprehensive analysis of protein-protein interactions in saccharomyces cerevisiae. *Nature*, **403**, 623–627.

Vázquez, A. (2001). Statistics of citation networks. arXiv.org:cond-mat/0105031.

Vázquez, A., Flammini, A., Maritan, A., and Vespignani, A. (2003). Modeling of protein interaction networks. *ComPlexUs*, **1**, 38–44.

Vázquez, A., Pastor-Satorras, R., and Vespignani, A. (2002). Large-scale topological and dynamical properties of the internet. *Physical Review E*, **65**, 066130.

Volterra, V. (1962). *Opere Matematiche*, Volume 5. Accademia Nazionale dei Lincei, Rome (Italy).

Wagner, A. (2001). The yeast protein interaction network evolves rapidly and contains few redundant duplicate genes. *Molecular Biological Evolution*, **18**, 1283–1292.

Warren, P.H. (1989). Spatial and temporal variation in the structure of a freshwater food web. *Oikos*, **55**, 299–311.

Wasserman, S. and Faust, K. (1994). *Social Network Analysis.* Cambridge University Press, Cambridge.

Watson, J.D. and Crick, F.H.C. (1953). Molecular structure of nucleic acids: A structure for deoxyribose nucleic acid. *Nature*, **171**, 737–738.

Watts, D.J. (1999). *Small-worlds: The Dynamics of Networks between Order and Randomness.* Princeton University Press, Princeton, NJ (USA).

Watts, D.J. and Strogatz, S.H. (1998). Collective dynamics of "small-world" networks. *Nature*, **393**, 440–442.

West, D.B. (2001). *Introduction to Graph Theory.* Prentice Hall, Upple Saddle River, NJ. 2-nd edition.

West, G.B., Brown, J.H., and Enquist, B.J. (1999). The fourth dimension of life: fractal geometry and allometric scaling of organisms. *Science*, **284**, 1677–1679.

White, H.D., Wellmann, B., and Nazer, N. (2004). Does citation reflect social structure? longitudinal evidence from the 'globenet' interdisciplinary research group. *Journal of the American Society for Information Science and Technology*, **55**, 111–126.

Wilkinson, D. and Willemsen, J.F. (1983). Invasion percolation: a new form of percolation theory. *Journal of Physics A*, **16**, 3365–3376.

Williams, R.J., Berlow, E.L., Dunne, J.A., Barabási, A.-L., and Martinez, N.D. (2002). Two degrees of separation in complex food webs. *Proceedings of the National Academy of Sciences*, **99**, 12913–12916.

Willis, J.C. (1922). *Age and Area.* Cambridge University Press, Cambridge.

Willis, J.C. and Yule, G.U. (1922). Some statistics of evolution and geographical distributions in plants and animals and their significance. *Nature*, **109**, 177–179.

Witten, T.A. and Sander, L.M. (1981). Diffusion-limited aggregation. *Physical Review Letters*, **47**, 1400–1403.

Yook, S.H., Jeong, H., Barabási, A.-L., and Tu, Y. (2001). Weighted evolving networks. *Physical Review Letters*, **86**, 5835–5838.

Yule, G. (1925). A mathematical theory of evolution based on the conclusions of Dr. J. C. Willis. F.R.S. *Phylosophical Transactions of the Royal Society of London*, **213**, 21–87.

Zanette, D. and Manrubia, S.C. (1997). Role of intermittency in urban development: A model of large-scale city formation. *Physical Review Letters*, **79**, 523–526.

Zipf, G.K. (1949). *Human Behavior and the Principle of Least Effort. An Introduction to Human Ecology.* Addison-Wesley, Reading MA.

Zlatić, V., Božičević, M., Štefančić, H., and Domazet, M. (2006).

Wikipedias: Collaborative web-based encyclopedias as complex networks. *Physical Review E*, **74**, 016115.

Zuckerman, H. (1977). *Scientific Elite: Nobel Laureates in the United States.* Free Press, New York.

Index